Taking Responsibility for Climate Change

Säde Hormio

Taking Responsibility for Climate Change

palgrave
macmillan

Säde Hormio
Practical Philosophy
University of Helsinki
Helsinki, Finland

Institute for Futures Studies
Stockholm, Sweden

ISBN 978-3-031-51752-5 ISBN 978-3-031-51753-2 (eBook)
https://doi.org/10.1007/978-3-031-51753-2

Cover image credit: RapidEye/Getty Images

This Palgrave Macmillan imprint is published by the registered company Springer Nature Switzerland AG.
The registered company address is: Gewerbestrasse 11, 6330 Cham, Switzerland

Paper in this product is recyclable.

To my future generation, Aaron and Kaius

Acknowledgements

Climate change presents an urgent and incredibly complex challenge for responsibility. In this book, I focus on the inherently collective and systemic nature of the problem, and how individual and collective responsibilities are tightly interwoven when it comes to addressing it.

I am very grateful for the following two grants for making the book possible: the European Union's Horizon 2020 research and innovation programme under the Marie Skłodowska-Curie grant agreement number GAP-839448 and Riksbankens Jubileumsfond grant number P22-0662.

For various reasons, the writing process was characterised by periods of intense work, followed by long breaks. The book has thus taken shape over several years. It has been helpful to have time to think through the multifaceted questions, although I am under no illusion that I have come up with all the answers.

The first versions of most of the chapters were written during my Marie Skłodowska-Curie Global Fellowship. My research project *Complicity: Individual Responsibility in Collective Contexts* (CIRICC) ran from August 2019 to July 2021. During the first year of the project, I was a Visiting Scholar at the University of California, Berkeley, where my work benefited from discussions with my sponsor, Christopher Kutz.

The manuscript took its final shape while I was a researcher in the *Ethics of Coordination* project at the Institute for Futures Studies (IFFS) in Stockholm, led by Krister Bykvist from Stockholm University. The camaraderie within the project research team has been helpful when thinking through questions of collective climate responsibility.

The manuscript has benefited from the generous, critical, and helpful comments from the three anonymous reviewers. I hope that I have done justice to all their suggestions in this final version. I am also thankful for the comments received at various conferences and workshops, where I have presented portions of the argument over the past few years.

The early version of Chap. 1 was given at the *American Philosophical Association (APA) Eastern Division Meeting*, held online in January 2021. A very rough first draft of Chap. 2 was presented in the *Climate Change Law and Policy: New Frontiers* workshop at Berkeley Law in February 2020, which I co-organised together with Hisayo Maitani Murakoshi and Gina Jeehyun Choi. A later version of the chapter was presented at the *Social Ontology in Political Philosophy* workshop at Bilkent University and the *Workshop on Agency* at the Helsinki Collegium for Advanced Studies, both in May 2023, as well as at the *Social Ontology 2023* conference in Stockholm in August that year. In addition, I received helpful suggestions for literature from Henry Shue, for which I am thankful.

Chapter 3 was presented at the *2020 APA Central Division Meeting* as part of the ISEE group session in Chicago in February 2020, and a later version at the *Ethics of Coordination* project workshop in March 2023. The first version of Chap. 4 was presented at *Small Acts, Big Harms: Workshop on Individual Responsibility for Collectively Caused Outcomes* in June 2021, which was the closing event for my CIRICC project and a good example of how high-quality philosophical debates can take place online as well. A final version was presented at the *First International Workshop on Coordination Ethics and Climate Change*, held at the IFFS in September 2023.

Chapter 5 took shape after I gave a talk on the central idea at the *IFFS Research Seminar* in December 2022. Chapter 6 was presented at the *Reasons and Rationality: Then and Now* workshop at Stanford University in January 2020, which I co-organised together with Oda Tvedt and Josiah Ober. I also presented the main argument at the IFFS online workshop *Empirical and Theoretical Perspectives on Green Growth and Climate Change* in April 2021, and at the *2022 Climate Futures Workshop: Layers of Governance*, held online in June 2022. Patricia Marino gave useful feedback on an early version. I also got valuable feedback on the chapter from Lukas Beck and Martin Kowarsch while I was finalising the manuscript.

In addition to the above, I presented portions of the book at the *Social Ontology 2018* conference at Tufts University, Boston, in August 2018, and in October 2020 at the *Reimagine the Future* discussion series, organised online by the Princeton University Energy Association.

I wish to thank the audience members for their invaluable questions and comments at all these events.

This book shares its overall argument about the many facets of climate change responsibility with my PhD thesis, *Marginal Participation, Complicity, and Agnotology: What Climate Change Can Teach Us About Individual and Collective Responsibility*. I defended it in December 2017 with John Broome as my opponent, from whom I received very generous and valuable feedback. While I tried out early versions of the arguments in the thesis, I did not want to publish it as it was. Instead, I wanted to make it into a proper book, so over the following years I re-worked the arguments. While some original parts of the text remain, I ended up writing the chapters afresh, although some of them naturally still share much of the literature and tone.

I would also like to thank my editor, Amy Invernizzi, who has continued to impress me with her dedicated professionalism from the first time she contacted me to suggest that I write a book, until the final stages of preparing the manuscript.

Last but not least, I want to thank Kaius and Aaron for being so patient with me while I worked towards finishing the book. This is for you.

Contents

About the Author

Säde Hormio is an Academy Research Fellow in Practical Philosophy at the University of Helsinki. She is also affiliated with the Institute for Futures Studies in Stockholm. While writing the book, Hormio was a Marie Skłodowska-Curie Global Fellow. She has published articles on collective responsibility, social epistemology, and climate ethics in journals such as *Wiley Interdisciplinary Reviews: Climate Change* and *Episteme*. She has also co-edited two volumes. This is her first monograph.

CHAPTER 1

Introduction to Collective and Shared Responsibility for Climate Change

Outline of the Main Arguments

Although the basic mechanism is simple—we burn too much fossil fuel—the result, climate change, is a very complex phenomenon. The science is complex, the implications are complex, and the number of agents involved in creating the harm is simply vast.

The starting position for this book is that not only states and international bodies, like the United Nations, have a responsibility to take action to mitigate climate change. Other collective agents, such as corporations, need to also be considered and to come onboard. Many kinds of collective agents can play various important roles in finding solutions. Furthermore, the book argues that climate change is not solely a problem for collective agents, but also for individuals, as they are members of collectives and groups of several kinds. Therefore, framing climate change responsibility exclusively from either the collective or the individual perspective leaves out something crucial: how we all are influenced by the collectives we belong to and how, in turn, collectives are influenced by individuals. The focus of the book is on areas of climate change responsibility that are often left out of the picture or get too little attention.

The main claim of the book is that collective responsibility must be construed widely to grasp the scope of climate change responsibility. Climate change requires an all-hands-on-deck approach at this point in time. We need collective agents of all stripes to participate in changing the

S. Hormio, *Taking Responsibility for Climate Change*,
https://doi.org/10.1007/978-3-031-51753-2_1

structures that lead to excessive greenhouse gas (GHG) emissions. This includes corporations and other companies, regional and local governments and their organisations, cities, banks, pension funds, non-governmental organisations, art and culture establishments, universities and other educational institutions, sports associations, and so on and so forth. Large-scale collective action is the key to the enormous scope and complexity of the challenge facing us, but not all of that action needs to be centrally coordinated. This is the topic of Chap. 2, in which my discussion centres on the responsibility of collective agents.

Throughout, my focus is on mitigating climate change; that is, reducing its harmful effects. The reason for my focus on mitigation rather than climate change adaptation or compensation is not that the latter two are not important (they are). The focus is on mitigation because I am interested in the unquestionably vast collective nature of the problem and what this means for responsibility. When any agent reduces their GHG emissions, the benefits fall on all. This is not necessarily the case with adaptation and compensation, which can take the shape of more regional and local efforts. Likewise, the harm caused by failure to mitigate falls on all agents, albeit not equally, something that will be discussed over the course of the book.

The second main claim in the book is that although collective agents should be at the forefront of mitigation efforts, their responsibility should not be discussed in isolation of the responsibility of their members. Organised collectives, such as universities, businesses, non-governmental organisations, or art and sports associations, have rules and decision-making procedures in place. They also have a core of some kind that can be identified as representing the collective, for example, executives and a board of trustees. Importantly, these collective actors come with differentiated roles. Individuals can share responsibility for collectively caused harm through their membership in collective agents. For example, employees of companies operating in fossil fuel intensive fields can be implicated as members of collective agents with high emissions, or citizens of industrialised nations as members of nation-states that could do something to address the issue. How we act in our various social roles is important, and we should not forget responsibilities of collective agents in getting rid of structural harms.

Shared responsibility as members of collective agents can be our key responsibility when it comes to climate change (depending on our circumstances and characteristics). Collective agents are organised to act as a collective entity, with decision-making mechanisms in place and often resources at their disposal. Members of any collective agent are the key to how that collective agent acts, although there is of course a lot of variation between collective agents in membership size and constrains brought by collective procedures. Still, shared responsibility as members is in many cases the best way an individual can try to make a difference to climate change. Collective entities should step up to meet the urgent need to mitigate climate change, and their members must push them to do so in the absence of effective legislation. Shared responsibility as members (individuals *qua* members of collective agents) is looked at more detail in Chap. 3.

In addition to shared responsibility as members, there are two other possible sources of moral responsibility for individuals in relation to climate change harms: shared responsibility as constituents (individuals *qua* constituents of unorganised collectives) and direct responsibility as super-polluters (individuals *qua* individuals). Organised, agential collectives are usually in a much better position to act than unorganised collectives. Saying that, shared responsibility as constituents can sometimes be an important factor too, even decisive, especially in offering support for grassroots political movements. Chapter 4 discusses how when we become aware of our shared social orientations, we can start to enact change in social norms and behaviours. This can make taking mitigation action more effective and also less socially costly for individuals. Chapter 5 discusses why in climate ethics it is important not to overlook how the consumption of a relatively small group of people is driving up the global emissions and how carbon inequality should impact climate mitigation responsibilities. When it comes to climate harms, individual direct responsibility is limited to wealthy individuals and their luxury emissions: to not to increase the probable risk of serious harm—deprivation of fundamental capabilities—to other people, at least as long as we can do so at a less than significant cost to ourselves. Offsetting is not a reliable way to meet this responsibility. While super-polluters have direct individual responsibilities, these are rarely prior to their shared responsibilities as members of organised collectives, in which they often occupy key positions.

But why should theoretical arguments about normative issues matter when we have a real-life climate crisis on our hands? I will argue in Chap. 6 that ethical arguments have a significant role in setting climate policy: they can highlight what values are at stake, inspire new horizons of thought, and help ground normative arguments in public deliberations.

Climate ethics matters in the real world because climate policy is about values. Value choices are not subject to optimisation and other formulaic solutions. These value choices should be debated openly to get climate policies right. Not only this, but ethical theories can also help to guide individual moral deliberations and help us to see the importance of working with others.

Before getting to these main arguments of the book, in this chapter I will first go through some background materials related to the urgency of the climate threat and the structural nature of the problem, as well as explain what collective agency is. I will also elaborate on the concept of responsibility that I am mainly operating with. These should help to clarify the discussions to follow.

Urgency

With climate change, the clock is ticking, and it has been for a while. Each year of delayed action is likely to result in a more chaotic climate and all the harms that it will bring. Higher global average temperature challenges the basic building blocks of a decent life, such as adequate food production and safe living environment. Increasingly unpredictable weather patterns bring on things like amplified droughts and flooding, and extreme weather events, such as stronger hurricanes and cyclones, will make it not only harder to grow crops but will also turn previously hospitable areas into unhabitable ones. Such changes are likely to result in increased conflicts, which will bring more suffering and harm. If we are to mitigate climate change to a level to which societies can still adapt to, urgent structural changes are called for. Above all else, the world needs a rapid phase out of its reliance on fossil fuels.

Ideally, humans need to not let the global average temperature rise more than 1.5 °C above pre-industrial times (IPCC 2018).[1] To achieve this, the world has to reduce emissions by 45% from 2010 levels by 2030 and to achieve net zero greenhouse gas emissions by 2050. This temperature is an aspirational target adopted as part of the political process leading to the Paris Agreement in 2015, and while it does not represent an ultimate threshold, science backs up the importance of not letting the global average temperature rise much further than it. This is because beyond this

[1] 1 °C was passed in 2015, and we are currently on course to reach 1.5 °C by the early 2030s.

point, the risk of extreme weather events—such as heat waves and droughts—will be considerably higher (IPCC 2018). Many argue that we are already too late to meet the 1.5 °C target. Indeed, the projections for where the world's emissions are heading, even if the current pledges and policies are all actioned, translates to at least 2.4–2.8 °C increase by the end of the century (UNEP 2022).

Tipping points provide additional uncertainty about projections and the risks involved in them. In climate science, *tipping point* is used to refer to critical thresholds that once crossed will likely lead to large-scale changes in the climate system, which are often irreversible. Examples include the melting of permafrost or the collapse of ice sheets. Crossing tipping points can lead to feedback mechanisms. Take the melting of ice sheets, for example. Ice is white, and white absorbs less heat than dark colours do (this is why you will find many buildings and roofs painted in light colours in countries with hot temperatures). When sea ice melts, it reveals the dark ocean below it. These dark waters absorb more solar heat, making the sea water warmer, which results in ever-faster melting of sea ice.

A lot of time has been wasted. Scientific consensus about the scale of the climate change threat first emerged around 1990, with the completion of the First Assessment Report of the Intergovernmental Panel on Climate Change (IPCC). While there is uncertainty over what tipping points would bring the climate system closer to collapsing (and if indeed we have already passed some), what is certain is that we are already perilously close to reaching 1.5 °C, that is, the threshold projected relatively safe by the IPCC. Therefore, whatever future emissions there are to be, restricting them is a matter of great urgency. The latest IPCC report (2022) states that the cumulative net CO_2 emissions of the last decade (2010–2019) are about the same size as the remaining carbon budget for keeping warming to 1.5 °C (medium confidence). So, the clock really is ticking. What is even more worrying is that the world is emitting CO_2 at an accelerated rate: 62% of total cumulative CO_2 emissions from 1850 to 2019 has occurred since 1970, about 43% since 1990, and about 17% since 2010 (IPCC 2022).

There is no magical technological fix coming to save us and take away the need to change our energy structures. Scientists have raised the alarm about the risks and potential side-effects of suggested geoengineering methods, such as solar geoengineering (Biermann et al. 2022). Apart from the high risks involved, geoengineering solutions will not be factored into the timeframe and on the scale required (Tavoni et al. 2012). This is not to say that we do not need new technologies to fight climate change. Carbon dioxide removal will be required to stabilise global warming to temperatures that will not be catastrophic to humans. The technology is needed both to achieve carbon neutrality in infrastructure going forward and to reduce the amount of carbon already in the atmosphere. However, the important role of CO_2 removal technologies in helping to keep within agreed global carbon budgets (i.e., how much carbon there is in total in the atmosphere) should not be confused with what Henry Shue (2021, 90) has called 'climate dreaming', the dangerous idea that there is not an urgent need to reduce emissions now, because some new technological inventions will come and save us. Humans will not be able to erase the effects of past emissions fully, because even if we could develop technologies for truly large-scale removal of carbon from the atmosphere, by that time we might have already passed tipping points with irreversible results. Or as Shue (ibid., 111) puts it: "*Temporary changes can produce permanent effects*" (italics in the original). As the Paris climate goals are slipping further away, with no credible pathway to 1.5 °C in place, the only way to avoid an accelerating climate disaster is an urgent system-wide transformation (UNEP 2022). Even if we cannot reach the 1.5 °C target, or the 2 °C target, every fraction of a degree still matters (ibid.).

Structural Aspects of the Climate Problem

Even if there was some technological invention forthcoming to fix climate change (which is not plausible), we are at a significant risk of new "climate changes" emerging. This is because climate change results from different kinds of structural causes. By structural causes I refer to material infrastructure that locks emissions into certain paths for many decades, and societal structures and systems, such as economic systems, which impact our choices. To give one example, in the global economic system, environmental impacts are still mostly externalised and fall outside the markets. If the people who feel the impacts from environmental degradation first and hardest are not the same people who are making the decisions that result in those impacts, then the odds in the structure are stacked against them, which can lead to structural injustices. This is especially so if the people are also marginalised in global politics. While I will give examples of structural issues related to climate change throughout the book, in this section I will first discuss infrastructure and individual choices, and then focus on structural injustices. Bringing in these very different aspects of structures underscores the inherently collective nature of the climate change problem.

One reason why climate responsibility debates must consider the collective context is that so many of our choices are constrained by infrastructure. There has been extensive debate about the effects of a person's emissions and how these could be conceptualised as being inconsequential looked at in isolation from other emissions (e.g., Cripps 2013; Sandberg 2011; Sinnott-Armstrong 2005). The puzzle is what can be the source of individual responsibility when no one can mitigate climate change on their own, and the effects of an individual's emissions are perceived as harmless. However, as the natural carrying capacity of the atmosphere has already been exceeded, all greenhouse gas emissions are now expected to cause harm. This is due to the physical disposition of the climate change phenomenon. The cumulative nature of the emissions and the really long timeframe for carbon staying in the atmosphere, combined with the butterfly effect, mean that all individual emissions are expected to cause harm in the long run (Broome 2019). Thus, all emissions matter. As John Broome (2019) has argued, they might not contribute to this specific storm or that drought, or to going over some particular tipping point, but over the decades and centuries that they spend in the atmosphere, they have countless opportunities to cause harm. The more greenhouse gas emissions there are in the atmosphere, the more unstable the climate will be, and the more extreme weather events there will be. While an individual cannot unilaterally stop climate change, their emissions can make the overall harm worse.

I find such an argument to be persuasive and plausible. Climate change is affected by what we all do and our choices are not inconsequential. That said, any account that focuses solely on the expected harm from our actions can easily become overly complicated and demanding, as we would have to know the expected utility and harm on innumerable things connected to our daily lives. In addition, we can exercise little choice over some of our emissions. This is because most of our emissions are tied to structures and many of our actions operate under collective level choices.

In cases in which the available infrastructure offers no options, an average individual has no small consumption choices to make. Consider the type of energy infrastructure that exists where you live. An individual might invest in solar panels for their own roof, if they can afford to do so, but the schools their children attend and the libraries they visit will be powered by coal power if that is the energy infrastructure of the city. The same goes for modes of transport. You can only take the metro, bus, or tram to work if an efficient enough public transport system is available. If public transport networks are lacking or inadequate, I might still have a choice of whether to take a car or bicycle (provided I have access to both a car and a bicycle). But even here available structures are likely to affect my behaviour: how safe is it to ride a bicycle where I live? For example, are there separate cycle lanes? In cases in which the infrastructure shapes our consumption habits, there is no direct small choice, that is, a choice that is entirely up to the individual (or if there is, it can be unfeasibly prohibitive: drive your car to work or quit). This does not mean that individuals have no options to act. They can work with others in the political arena to try to bring about structural changes. My point is only that their consumption choices are limited by structures, and such cases limit individual responsibility to a great extent, although it does not have to block it completely, as I will argue in Chap. 4.

Even in cases in which there is a direct small choice, the structures could block there being a direct small effect. For example, daily meat consumption has become a social norm in many countries due to the availability of cheap meat, made possible by mass farming methods and centralised slaughterhouses. This has major consequences for overall emissions, as the livestock sector is estimated to account for somewhere between 11% and 19% of all human-induced yearly emissions (FAO 2013; Xu et al. 2021, see also GLEAM data from 2015).[2] Beef and milk account for the majority of emissions, with beef production accounting for 41% of the sector's emissions and milk for 19%.[3] What makes the figures even more important is that approximately 44% of the sector's emissions are methane, the strongest greenhouse gas.[4] While methane lasts for only

[2] The FAO maintains the Global Livestock Environmental Assessment Model (GLEAM) on the impact of livestock on the environment, available at https://foodandagricultureorganization.shinyapps.io/GLEAMV3_Public. The latest available global data is from 2015, estimating livestock to account for 11.2% of total global GHG emissions.

[3] The emissions linked to the meat industry comprise mainly of feed production and processing (45%), enteric fermentation, i.e., outputs of methane from the animal's digestion process (39%), and manure decomposition (10%). The remainder is attributable to the processing and transportation of animal products (FAO 2013).

[4] Cattle and other ruminant animals (such as buffalo, sheep, and goats) produce methane as part of their digestive process. Microbial fermentation breaks down carbohydrates in their stomach (rumen) into simple molecules that can then be digested by the animals. Methane is a by-product of this process. Poorly digestible, fibrous food causes higher methane emissions, so what feed the cattle get is important also. Non-ruminant species, such as pigs (and humans), also produce methane during digestion but amounts are much lower by comparison. Methane is also released when faeces decompose. This occurs mostly when animal manure is processed and held in liquid form, such as in deep lagoons or holding tanks commonly used in industrial meat production (FAO 2013, 20).

about 12 years in the atmosphere, it warms the planet roughly 30 times more than CO_2, giving urgency to curbing methane emissions.

While daily meat consumption norms have increasingly been questioned, an individual going vegetarian is highly unlikely to reduce the methane emissions from a given farm (unless it is their farm), let alone from the meat industry (setting aside the possibility of influencing others) even if they have a direct small choice of whether to eat meat or not. This is because the outcome is not incremental: you do not slaughter 1/8th of a cow less (illustrating a made-up individual yearly beef consumption[5]), any more than you can raise 1/8th of a calf less. In other words, there is a methane-emitting unit, namely a cow, and there is a certain threshold of demand for more cows to be bred and slaughtered based on how much meat each cow can be turned into. Note that this example assumes that the individual changes their consumption patterns. A one-off direct small choice, such as choosing a vegetarian option in a restaurant over a beef option, has even less chance of having any effect on the methane emissions from livestock farming—in isolation from other such acts—as one meal represents 1/2000th of a cow.[6] Still we have good reasons to choose the vegetarian option, at least most of the time, or so I will argue. More specifically, I will argue in Chap. 4 that when we become aware of a normatively relevant shared social orientation, it can give us a reason to make (or refrain from making) small contributions to collective outcomes, regardless of the expected utility (or lack thereof) of the small contribution.[7] As part of our consumption falls under collective level structures, the choices

[5] It assumes that the individual eats around 25 kg of meat in a year, which translates into roughly one-eighth of a cow. You obtain about 195 kg (430 pounds) of retail cuts from a steer (a young, neutered male raised for beef) that weighs 454 kg (1000 pounds) on the hoof, according to an estimate by the Meat Inspection Services at the Food Safety Division of the Oklahoma Department of Agriculture Food and Forestry ("How Much Meat?", undated document available at www.oda.state.ok.us/food/fs-cowweight.pdf, accessed 28 May 2020). Beef consumption varies considerably between countries and individuals, but many of the yearly annual consumption figures per person are calculated based on the full weight of the animal, not on the retail cuts.

[6] Using the above figures and assuming a portion size of 100 g (3.5 oz), one portion at a restaurant is 0.05128% of a cow, i.e., roughly 1/2000th of a cow.

[7] Kagan (2011) argues that collective action problems can be solved straightforwardly by consequentialist reasoning, as individual acts might make a difference due to some threshold being crossed. For criticism of Kagan's expected utility solution to his chicken-eater's dilemma, see Nefsky (2011) and Lawford-Smith (2016).

and responsibility of individual consumer should not be debated in isolation from these.

So far I have discussed mainly infrastructure, but as I noted in the beginning of the section, structural issues linked to climate change are wider than this, and social structures can even lead to structural injustices.[8] In the words of Iris Marion Young (2011, 52), structural injustices are consequences of social-structural processes that "put large groups of persons under systematic threat of domination or deprivation of the means to develop and exercise their capacities". A general feature of structural injustices is that they are not necessarily traceable to the actions of individual agents, at least in an obvious way. They also need not be intended; rather, they are a side-effect of some other action. The primary structural injustice with climate change is that the worst impacts will be felt by many of the least economically affluent communities, despite them contributing to the problem the least (Eckersley 2016). The marginal social-structural position of these communities, both in the economy and in the state system, exposes them to most of the risks that are generated by the social structures, with few of the benefits. The added injustice is that due to these structures, these communities are placed in a particularly weak position to coordinate the changes necessary to reduce their vulnerability.

Many communities of Indigenous people are feeling the impact of climate change in their daily lives already, as they often live in ways that are in tune with weather patterns. Rapidly changing weather conditions have led to unpredictable seasons which pose a threat to many such communities. For example, increasingly severe and less predictable weather and shrinking ice packs are already impacting Inuit livelihoods by making it harder to utilise traditional knowledge (Inuit Circumpolar Council 2014). Climate change has made travelling difficult and dangerous, isolating communities, and cutting Inuits off from their traditional hunting lands. Not only do the changes brought about by climate change make it harder to hunt for food, but they also disrupt cultural traditions that are hundreds of years old and can affect the mental wellbeing of the people (Mercer 2018). Climate change can also pose a threat to the livelihood and culture of Indigenous communities when they are settled and high

[8] The remaining of this section draws from my 2021 APA blog post, *Individual Responsibility and the Structural Harm of Climate Change*, available at blog.apaonline.org/2021/02/22/individual-responsibility-and-the-structural-harm-of-climate-change, reprinted with permission of the American Philosophical Association.

mobility is not one of their adaptive strategies. Consider the Sámi people in Finland, Norway, Russia, and Sweden, who have a long tradition of reindeer herding. Temperatures have been rising faster in Arctic regions than the global average and the resulting milder winters have changed the snow conditions. While snow used to last, it now melts and refreezes, preventing reindeer from smelling and digging for lichen, their favourite food, a symbiosis of fungus and alga, now trapped under hard snow and ice (Narang 2018). Another threat to reindeer herding is presented by the new parasites and diseases that are expanding north as temperatures get warmer.

However, it is not the case that Indigenous people are more at risk of climate change simply by choosing to live in certain places and in certain ways. Instead, the Indigenous climate justice movement points out how ongoing, cyclical colonialism is often a major reason for the current vulnerability of these communities.[9] If we do not include an assessment of historical injustices and how they have led to increased vulnerability to climate change harms, the reason that Indigenous peoples face greater risks seems only a matter of chance. This position is labelled by Kyle Powys Whyte (2017) as "The Bad Luck View". Legacies of colonialism include multi-faceted problems such as marginalisation and poverty, and the resulting socio-economic conditions are far from ideal for absorbing and withstanding climate change impacts. Still, climate change impacts should not be seen as a new problem that exacerbates old problems. Rather, the problems are related and form a continuum: the same institutions that facilitate carbon-intensive economic activities also interfere with Indigenous peoples' capacity to adapt to adverse climate impacts. In the words of Lisa Herzog (2017), employing the notion of a structural injustice locates the focus on the "social *positions* that individuals can end up in, emphasising that these should neither be understood as the result of

[9] It is important to note that the marginalisation of Indigenous people in politics is not due to passivity. In fact, the voices of Indigenous people are among the more audible in the global climate justice social movement. There are many grassroots campaigning organisations that actively campaign on climate change issues, like Idle No More and Land is Life. Another example is how many delegates from Indigenous communities attended the 2015 Conference of the Parties in Paris, although references to their rights were cut out from the binding portions of the agreement (Paquette 2016). In addition to Indigenous representatives at the UN's climate change talks, the communities also raise the issue through channels such as the Saami Council.

personal failure nor as caused by unhappy circumstances for which no one is responsible."

Although I will not be focusing on Indigenous people in the following chapters, their situation highlights how climate change is inescapably a structural issue. Emissions that cause anthropogenic climate change take place within systems. They are tied to both infrastructure and social structures, including capitalist structures. Responsibility for structures is inherently collective and cuts across generations. Although those alive today have inherited the infrastructure that is reliant on fossil fuels from past generations—partly created when the threat was not yet known or fully appreciated—this is not the whole story. The actions of our ancestors have raised global average temperatures, but the current generations have upheld, re-created, and expanded the problematic infrastructure and social practices. This is why we need to take a structural view whenever we discuss climate change responsibility.

Before discussing responsibility, though, I will first take a look at collective agents. If there is one thing that all climate ethicists seem to agree on, it is that it would be futile and insufficient to focus on the responsibilities of individuals to combat climate change, because an effective remedy must come from governments and other large collective agents. As Tracy Isaacs (2011, 140) puts it, with large-scale collectively caused harms like climate change, solutions focusing on individuals "are akin to attempting to put out a blazing house with a toy water pistol". Government policies can affect individual behaviour on a large scale by means such as taxation and laws which offer incentives and deterrents. The actions of other collective agents matter as well, but before we discuss their responsibility, I find it useful to explore what collective agents are first.

Collective Agency

Collective agents are organised collectives, meaning that the collective has an organisational structure which lays out their internal policies, procedures, and rules, giving each collective agent their own organisational character. Within this structure, we find embedded the normatively bound roles of its members (Ritchie 2020). However, it is not just the official rules and policies that give collective agents their unique flavour, as they also have their particular (corporate) culture, which is shaped by both formal and informal rules and interactions. With 'formal', I refer to both official, written down rules and policies, but also to the wider requirements of behaviour within the collective. After all, not everything contained within the corporate culture can be written down, and many of the formal expectations are learnt only after we become members of that collective and get accustomed to its social norms and ways of working. What unifies such collectives is not necessarily a shared or joint intention to do something together, like go for a walk, move a piano, or paint a house (c.f. Bratman 1992; Gilbert 1990; Tuomela 1989). To be sure, the collective agent has goals and plans, but the individual members do not necessarily share them. Rather, they participate in bringing about the collective's goals and try to realise its plans for other reasons, which could vary widely between members. Importantly, the decisions of collective agents are independent of the intentions of individual members (although they can still be influenced by them, as I will argue).

Even though their organised structures set collective agents (Chaps. 2 and 3) apart from unorganised collectives (Chap. 4), collective agents differ greatly in respect of the degree to which they are organised. At one end of the spectrum we have formal policies, some legally binding, with clear hierarchies and procedures for coming up with decisions. Organisations and institutions are rule-guided, born out of the human capacity for planning agency (Bratman 2022). Legally recognised collective agents include nation-states, governments, non-governmental organisations, corporations, universities, orchestras and theatre groups, educational institutions, political parties, and sports clubs and professional sports teams, among other examples. While all these collectives have a structure, they differ in terms of how much the law regulates their activities. For example, collectives such as corporations and political parties must conform to regulations to a much wider extent than theatre groups or amateur sports clubs. Naturally, the latter are also subject to certain laws and regulations, but their activities are not usually affected by legislation to the same degree, nor are they given the same protections. (For example, corporations are given corporate personhood under US law.) Of course, a highly organised collective can also be without any legal protection, like a criminal organisation. Some collective agents write the laws for other collective agents, so the policies and procedures of one can affect others. There are many power relations between collective agents that affect questions of responsibility.

Collective agents are often discussed in the literature as corporate agents. A lot of this literature focuses on the formal decision procedures of corporations and other such highly organised collective agents. Peter A. French (1979) influentially pointed out that the internal decision structure of corporate agents must include two components: the rules that cover its procedures and policies (a corporate policy statement), and the corporation's internal roles and managerial lines (an organisational flow chart delineating responsibilities). When decisions fall within the parameters of the corporate's internal decision structure, the acts of individual members are incorporated into a collective act, which is recognised as the act of the corporation. Christian List and Philip Pettit (2011) have focused on voting procedures and how the official positions of corporate agents do not necessarily need to align with the individual positions of any of their members.

At the other end of the spectrum, towards the low degree of organisation, we have organised collectives that are identifiable only by informal policies, procedures, and hierarchies. A group of workers who are in a habit of going for a drink after work on Fridays would fall into such category. They are not an unorganised group of people. Instead, they have existing relations between them and a collective goal (going out together for a drink). These are the kind of groups that are discussed in the collective intentionality literature, as the collective goal can easily be cashed out in terms of a shared or a joint intention (to go out for a drink together). There are some informal rules and policies: the drinks are usually on Fridays, the people involved tend to work at the same place, and so on, so it could be described as a collective agent only in the weakest sense of the word. None of these rules and policies are formal requirements, but they could go out on a different night and so on. However, discussing the climate change responsibilities of such collectives is not very meaningful, so I will leave out such organised collectives with no formal rules from the category of collective agents. (Chapter 4 covers shared responsibility as constituents of unorganised collectives.)

I posit that there needs to be some minimal level of formality to the structure of a collective agent, although the structure does not need to refer to any legal manners and it can be mostly informal. An example would be a local Lord of the Rings fan club, which meets roughly once a month to discuss all matters Tolkien, but whose only informal rule or procedure is that all members must be able to converse in at least one of the Elvish languages. Note that it is a separate issue if such minimally formal collective agents are interesting in terms of climate change responsibility. That will depend on what their activities are. As the GHG emissions from monthly discussions held in Sindarin or Quenya are minimal, it is highly unlikely that such a collective agent is relevant for climate change responsibility, but a local aviation club certainly would be, and even a cooking club might be.[10] Of course, many collective agents fall somewhere between the extremes on the spectrum (that is, the scale of the degree of organisation), including families, hobby groups, loose associations revolving around shared interests, and the like. In this book, I am mostly interested in the collective agents that are on the more organised end of the spectrum.

There can also exist organised collectives that have both membership structure and decision-making procedures in place, but that are completely inactive or rarely do anything. Think of a secret society set up by a group of schoolchildren that met only a few times before the members lost interest in the game. Non-active organised collectives could also have a legal status, like a dormant company, registered but not operating, existing in name only. I am not interested in such collectives, so the debate is limited to collective agents with ongoing, persistent activities.

At the beginning of this section, I wrote that each collective agent has its own organisational character not just through the official rules and policies, but also through their particular (corporate) culture. Importantly, this is shaped by both formal and informal rules and interactions. It would be foolish to deduce merely from a robust ethics code that a collective agent is indeed ethical, just as it would be naïve to think that a green mission statement featured prominently on a corporate agent's website makes them automatically green. The true prominence of such issues is revealed only by inspecting the corporate culture, its ethos. Individual members affect the collective ethos and are in turn affected by it (Hormio forthcoming). The culture reveals what really goes on and what values get promoted in its daily operations. As Kendy Hess (2014) has pointed out, the decisions of collective agents are shaped by not just the formal decision mechanisms, but also more informal interactions. Importantly, members of a collective agent are often likely to be able to influence the development of corporate decisions and beliefs (Hess 2018).

[10] This is assuming that the local fan club's members live nearby and can come to the gatherings on foot, bicycle, short-distance public transport, etc. If they all lived far away from each other and would have to make long journeys to attend meetings, then there might be a responsibility to reduce the emissions caused by having at least some of them virtually. Note also that here I am discussing responsibility *qua* the collective agent. It is a separate issue if the members should engage in other forms of collective action, like climate activism, in addition to (or instead of) whatever the collective agent is involved in.

The responsibility of collective agents is discussed in Chap. 2. It is uncontroversial to assign the main responsibility to dealing with climate change to collective agents such as states (e.g., Broome 2012; Shue 2021; Vanderheiden 2008; for an overview see Hormio 2023). However, as an effective collective scheme for dealing with climate change is still lacking, should individuals do something in the face of it? Moreover, how should we conceptualise the responsibility of an individual to a collectively caused structural harm that they did not intend to bring about? I will next take a closer look at the kind of responsibility that the book is concerned with.

Responsibility

Most of our emissions take place within structures. Individual responsibility should not be discussed in isolation from the collective settings that we are all embedded in. The collective setting affects the actions and decisions that are taken within, for, and by the collective agent in many ways. This means that often the actions and omissions of individual members cannot be separated from the incorporation acts and the collective processes that they are a part of. Although the focus of this book is on social and political philosophy, which explore the sphere in which collective agents operate, bringing in individual responsibility takes us also to the realm of moral philosophy. Within this realm, 'responsibility' has many associations (Williams 1985). My focus is not so much on what is the exact duty or an obligation of a particular agent (as that is always contextual), or how blameworthy they are, although questions about past actions and failings are also relevant for climate justice. Instead, I am mainly interested in what agents, both individual and collective, have good reasons to do in light of climate change, and what they arguably should do from now on.

My concern is thus mostly about *how* individuals should act, rather than trying to justify *whether* they should act. Climate change presents such an urgent and large-scale threat that it seems more fruitful to concentrate on the how, instead of trying to persuade those who are still unconvinced about the need for action. (Saying that, there is a lot of misinformation around climate science, mostly manufactured by those with vested interests in fossil fuels to make people uncertain about the need to take urgent action. This is discussed in the next chapter.)

The concept of responsibility that I am chiefly operating with throughout the book is an active stance. As the title suggests, I am interested in the concept of *taking* responsibility for one's actions and their impacts, including those that were not intended and have been caused with other agents. This does not translate to that we should somehow blame ourselves or feel guilty about our part. Such feelings are not necessary for taking responsibility for something. When I bump into you on a busy street as you run into my path and fall down, I can choose to take responsibility and apologise, all without a morsel of guilt or remorse. My feelings here are closer to agent-regret for having been implicated in a harm (Williams 1981). It is also commonplace to take responsibility for actions done by others. Think of a parent of a child who has broken a window, or the boss of an employee who has messed up. Life also constantly presents us with situations in which we have conflicting duties and pressures, where there is no perfect way to choose to behave, but only better and worse options. It is up to us how we choose to act.

I subscribe to the idea presented by Paulina Sliwa (forthcoming) that to take responsibility, we do not need to feel guilty or remorseful, but we need affective engagement with the outcome in a form of sympathy or regret, for example. According to her, taking responsibility for one's actions is a matter of owning one's normative footprint, that is, accepting the consequences of one's actions by acknowledging their normative consequences. This has cognitive, conative, and affective components. The cognitive involves an understanding of the normative consequences, the conative component involves motivation to comply with the duties incurred, while the affective requirement is about being responsive to the changes in "feeling-duties" in play. This means that while guilt and remorse are not necessary, according to Sliwa, I need to have certain care and concern for the other party and the harm they have suffered. The parent of the child who has broken a window might feel genuinely sorry for the trouble that the neighbour has to go through to fix their window, which goes beyond mere acceptance of financial liability to cover the costs. Similarly, the boss of an employee who has messed up could feel distressed on behalf of the customer who has had to bear the brunt of the mess caused by their employee. However, they do not need to feel any guilt over this, and it would not make much sense either. Sliwa underlines that taking responsibility for the wrongdoings of others is not a matter of somehow "taking on" their normative footprint, but rather about owning the consequences that their wrongdoing has for our *own* duties and obligations.

I suggest that this can also be applied to the relationship between collective agents and their members. When a member of a collective agent takes responsibility for harms caused by the collective agent, they are not taking on the normative footprint of that collective agent. It would never make sense for an individual to do so in any case. They could not: collectives can do things individuals cannot do alone and ought implies can. Rather, taking responsibility as a member of a collective agent is about owning the consequences that the harm caused by the collective has for one's own responsibilities as a member of that collective. The same logic applies to holding others responsible for their parts: it makes sense to do so even when we are not blaming them.

Another useful concept in teasing out the kind of responsibility that is the main focus of this book is that of formative moral criticism. Robin Zheng (2021) describes formative criticism as feedback aimed at improving an agent's performance in an ongoing way, in contrast to summative criticism that assess performance relative to some benchmark. She notes how it is hard to live ethically in a highly unjust world in which we all face structural constraint in our choices. Formative responses consider how our moral agency is constrained in important respects and can be thwarted by social forces. The feedback aims to improve individuals' actions and attitudes. In aspirational morality, an agent is ongoingly striving towards an ideal, so what is evaluated is not a particular act, but rather the ongoing activity of striving towards morally aspirational ideals. This is despite them being unrealisable. Feedback to others in the form of formative moral criticism is meant to enhance our moral aspirations. According to Zheng, it is to help us to try to learn, feel, and be motivated to take action, re-enforcing our efforts to do better. There is also research on how recognising personal responsibility can activate self-conscious emotions that motivate people to act in line with their moral standards (Brouwer et al. 2022). These could be blame, but equally well about sympathy, agent-regret or other such moral emotions. Formative moral criticism is thus not interested in assessing the quality of agents, but rather in redirecting their attention to the problem at hand, which often translates to joining with others to collectively transform social structures.

This fits with my goal in this book, which is not to assess who the "baddies" are (even when I discuss super-polluters in Chap. 5), but rather to try to focus on what is problematic about our structures and shared practices and what we could do together to correct the course. This does not take away the fact that sometimes we should also blame other agents (or ourselves) in the climate change case. For some of the most egregious examples, look for discussion in the next chapter on fossil fuel corporations and their enablers. But the point remains that it makes sense to talk about taking responsibility for something, even when there is no blame involved (rather, aspirational feedback). However, as the problems associated with the required changes are so vast and complex, climate solutions will also be complex and will need input from various sources. Often there will be no one clear solution, and that is why we need also participatory democracy (Chap. 6) to set the course. Philosophical and other normative arguments have their role to play in highlighting the different values at stake when making large-scale societal decisions and suggesting the best routes forward, but the actual course must be mapped out together.

Bringing in the contextual nature of morality and discussing the merits of participatory democracy could be seen to be in conflict with the moral arguments I make in the earlier chapters. Are the conclusions I reach true regardless of any democratic process they might or might not be part of, or should ethical arguments be subjected to democratic scrutiny? I find that the answer to this lies in the scope of the ethical argument and on how we understand the universality of moral principles. When we discuss societal matters and how to organise our lives together (including collective agents and their responsibilities within a society), I believe that there is an element of democratic scrutiny that is warranted for any theories put forward. In democracy, decisions are better if they can be justified to others. That they are decided together democratically has also its own value. Furthermore, there is also added epistemic value if participatory processes help us to remove blind spots within theories (see Chap. 6).

The aim of democracy as a process is not to find the truth but instead what is best for the whole. Ethical arguments operate on a different logic. In moral matters, we should always be willing to treat others in ways that we can justify to them, but this does not require democratic processes. Ethical theories can help us to clarify our thinking and suggest new routes, perhaps also to inspire us to do better. I believe that good ethical theories are the best attempts by their writer(s) at approaching a moral conclusion using the methods and rigorous tools of philosophy. Still, even the best theories are still always contextual to some degree, as an action that is good or permissible in one context can be bad and impermissible in another. This can be true either due to social conventions understood as established practices, or because different social conditions give different reasons for people to reject proposed principles (Scanlon 1998, 338–340). However, this does not mean that social factors and realities alone determine what is right or wrong, as there are certain judgements of right and wrong that hold anywhere anytime (ibid., 348). It is important to remember that there are some ethical principles that are fundamental and widely shared. The values attached to these do not usually change often or quickly, due to the principles being built on experience accumulated over centuries (Kowarsch and Edenhofer 2016). Certain judgements of right and wrong hold anywhere anytime, regardless of if they are recognised as valuable within political systems, including democratic ones.

Apart from concerns about the consequences of our actions (in this case, about the effects of our emissions), another equally important aspect of my account of responsibility involves becoming aware of the opportunities we have for enacting change, especially when we work together with others. This side of the responsibility coin is purely forward-looking and focuses on the abilities we can have to try to impact how things are going. The exact responsibilities of a given agent are always contextual, but we can formulate *pro tanto* reasons regarding climate change action, reasons that can lead to *pro tanto* duties, which should be taken seriously, but can be outweighed by other duties, depending on the circumstances.

Throughout, I am concerned with the intersection of individual and collective responsibility, as I believe it is where we find motivating reasons for action. My main focus is on mitigating future GHG emissions, which means both trying to avoid engaging in actions that increase emissions and to try to reduce the climate impacts of our existing activities. Correspondingly, most of the discussion is on what collectives and individuals across the board should do going forward. Still, as climate change impacts are already underway, there is also a need to repair and to provide compensation for damage already caused. This calls for debate on responsibility for past emissions, and this will be focused on the fossil fuel industry and states. Historical emissions can impact both the corrective question about how to divide the costs for the effects of unmitigated emissions and distributive question about how to divide the remaining carbon budget. Many of these issues link us to structures that the emissions take place in.

However, when collective solutions are not there yet, the spotlight of responsibility seeks individuals. If the attempt to delegate responsibility effectively to political institutions has failed, the responsibility falls back on citizens again (Gardiner 2011, 433). In other words, when there is no such delegation, the responsibilities fall back on individuals who are able to help (Stemplowska 2016). This is one of the main reasons why I want to highlight the aspect of individual responsibility that I believe has not received enough attention in the literature: individuals *qua* members of organised collectives. Shared responsibility *qua* members presses especially on those occupying key positions within key collective agents, as it is affected by your role within the collective. We should assess our roles and social positions, along with our personal circumstances and qualities, and see where we could make a difference, and then try to push for more carbon-neutral practices in ways that are most fitting for us. Climate saintliness is not the goal. Climate sanity will do.[11]

References

Biermann, Frank, Jeroen Oomen, Aarti Gupta, Saleem H. Ali, Ken Conca, Maarten A. Hajer, Prakash Kashwan, Louis J. Kotzé, Melissa Leach, Dirk Messner, Chukwumerije Okereke, Åsa Persson, Janez Potočnik, David Schlosberg, Michelle Scobie, and Stacy D. VanDeveer. 2022. Solar Geoengineering: The Case for an International Non-use Agreement. *WIREs Climate Change* 13 (3): e754.

Bratman, Michael E. 2022. *Shared and Institutional Agency: Toward a Planning Theory of Human Practical Organization.* New York: Oxford University Press.

———. 1992. Shared Cooperative Activity. *The Philosophical Review* 101 (2): 327–341.

Broome, John. 2012. *Climate Matters: Ethics in a Warming World.* New York: W. W. Norton.

———. 2019. Against Denialism. *The Monist* 102 (1): 110–129.

Brouwer, Claire, Jan Willem Bolderdijk, Gert Cornelissen, and Tim Kurz. 2022. Communication Strategies for Moral Rebels: How to Talk About Change in Order to Inspire Self-Efficacy in Others. *WIREs Climate Change* 13 (5): e781.

Cripps, Elizabeth. 2013. *Climate Change and the Moral Agent: Individual Duties in an Interdependent World.* Oxford: Oxford University Press.

Eckersley, Robyn. 2016. Responsibility for Climate Change as a Structural Injustice. In *The Oxford Handbook of Environmental Political Theory*, ed. Teena Gabrielson, Cheryl Hall, John M. Meyer, and David Schlosberg, 346–361. Oxford: Oxford University Press.

FAO. 2013. *Tackling Climate Change Through Livestock: A global assessment of emissions and mitigation opportunities.* Rome: Food and Agriculture Organization of the United Nations.

[11] I am paraphrasing Richard Heede, a researcher specialising in carbon extraction and emissions, who stated in an interview that "Climate saintliness is not required; climate sanity will do" when answering a question on what individuals should do (Horgan 2020).

French, Peter A. 1979. The Corporation as a Moral Person. *American Philosophical Quarterly* 16 (3): 207–215.
Gardiner, Stephen M. 2011. *A Perfect Moral Storm: The Ethical Tragedy of Climate Change*. New York: Oxford University Press.
Gilbert, Margaret. 1990. Walking Together: A Paradigmatic Social Phenomenon. *Midwest Studies in Philosophy* 15 (1): 1–14.
Herzog, Lisa. 2017. What Could Be Wrong with a Mortgage? Private Debt Markets from a Perspective of Structural Injustice. *Journal of Political Philosophy* 25: 411–434.
Hess, Kendy M. 2018. Who's Responsible? (It's Complicated.) Assigning Blame in the Wake of the Financial Crisis. *Midwest Studies In Philosophy* 42 (1): 133–155.
———. 2014. Because They Can: The Basis for the Moral Obligations of (Certain) Collectives. *Midwest Studies in Philosophy* 38: 203–221.
Horgan, John. 2020, June 2. Exposing the World's Biggest Carbon Emitters. *Scientific American*. https://blogs.scientificamerican.com/cross-check/exposing-the-worlds-biggest-carbon-emitters.
Hormio, Säde. 2023. Collective Responsibility for Climate Change. *Wiley Interdisciplinary Reviews Climate Change* 14 (4): e830.
———. forthcoming. Collective Agents as Moral Actors. In *Collective Responsibility: Perspectives on Political Philosophy from Social Ontology*, ed. Säde Hormio and Bill Wringe. Studies in the Philosophy of Sociality, Springer.
Inuit Circumpolar Council. 2014. *The Sea Ice Never Stops: Circumpolar Inuit Reflections on Sea Ice Use and Shipping in Inuit Nunaa*. Inuit Circumpolar Council.
IPCC. 2018. Summary for Policymakers. In *Global Warming of 1.5°C. An IPCC Special Report on the Impacts of Global Warming of 1.5°C Above Pre-industrial Levels and Related Global Greenhouse Gas Emission Pathways, in the Context of Strengthening the Global Response to the Threat of Climate Change, Sustainable Development, and Efforts to Eradicate Poverty*, ed. V. Masson-Delmotte, P. Zhai, H.-O. Pörtner, D. Roberts, J. Skea, P.R. Shukla, A. Pirani, W. Moufouma-Okia, C. Péan, R. Pidcock, S. Connors, J.B.R. Matthews, Y. Chen, X. Zhou, M.I. Gomis, E. Lonnoy, T. Maycock, M. Tignor, and T. Waterfield. Cambridge: Cambridge University Press.
———. 2022. Summary for Policymakers. In *Climate Change 2022: Mitigation of Climate Change. Contribution of Working Group III to the Sixth Assessment Report of the Intergovernmental Panel on Climate Change*, ed. P.R. Shukla, J. Skea, R. Slade, A. Al Khourdajie, R. van Diemen, D. McCollum, M. Pathak, S. Some, P. Vyas, R. Fradera, M. Belkacemi, A. Hasija, G. Lisboa, S. Luz, and J. Malley. Cambridge: Cambridge University Press.
Isaacs, Tracy. 2011. *Moral Responsibility in Collective Contexts*. New York: Oxford University Press.

Kagan, Shelly. 2011. Do I Make a Difference? *Philosophy & Public Affairs* 39 (2): 105–141.

Kowarsch, Martin, and Ottmar Edenhofer. 2016. Principles or Pathways? Improving the Contribution of Philosophical Ethics to Climate Policy. In *Climate Justice in a Non-Ideal World*, ed. Clare Heyward and Dominic Roser, 296–318. Oxford: Oxford University Press.

Lawford-Smith, Holly. 2016. Difference-Making and Individuals' Climate-Related Obligations. In *Climate Justice in a Non-Ideal World*, ed. Clare Heyward and Dominic Roser, 64–82. Oxford: Oxford University Press.

List, Christian, and Philip Pettit. 2011. *Group Agency: The Possibility, Design, and Status of Corporate Agents*. Oxford: Oxford University Press.

Mercer, Glen. 2018. 'Sea, ice, snow ... it's all changing': Inuit Struggle with Warming World. *The Guardian*, May 30.

Narang, Sonia. 2018. Finland's Reindeer-Herding Sámi Women Fight Climate Change. *The World*, March 7.

Nefsky, Julia. 2011. Consequentialism and the Problem of Collective Harm: A Reply to Kagan. *Philosophy & Public Affairs* 39: 364–395.

Paquette, Mitch. 2016. Indigenous Rights Cut from Paris Agreement: Why It Concerns Us All. *Intercontinental Cry*, January 13. intercontinentalcry.org/Indigenous-rights-cut-from-paris-agreement-why-it-concerns-us-all

Ritchie, Katherine. 2020. Minimal Cooperation and Group Role. In *Minimal Cooperation and Shared Agency*, ed. Anika Fiebich, 93–109. Studies in the Philosophy of Sociality: Springer.

Sandberg, Joakim. 2011. 'My Emissions Make No Difference': Climate Change and the Argument from Inconsequentialism. *Environmental Ethics* 33: 229–248.

Scanlon, T.M. 1998. *What We Owe to Each Other*. Cambridge, MA: The Belknap Press of Harvard University Press.

Shue, Henry. 2021. *The Pivotal Generation: Why We Have a Moral Responsibility to Slow Climate Change Right Now*. Princeton: Princeton University Press.

Sinnott-Armstrong, Walter. 2005. 'It's Not *My* Fault: Global Warming and Individual Moral Obligations. In *Perspectives on Climate Change: Science, Economics, Politics, Ethics*, ed. Walter Sinnott-Armstrong and Richard B. Howarth, 285–307. Oxford: Elsevier.

Sliwa, Paulina. forthcoming. Taking Responsibility. In *Normative Philosophy: Conversations in Moral, Legal, and Political Philosophy*, ed. Ruth Chang and Amia Srinivasan. Oxford University Press.

Stemplowska, Zofia. 2016. Doing More Than One's Fair Share. *Critical Review of International Social and Political Philosophy* 9: 591–608.

Tavoni, Massimo, Shoibal Chakravarty, and Robert Socolow. 2012. Safe *vs.* Fair: A Formidable Trade-off in Tackling Climate Change. *Sustainability* 4 (2): 210–226.

Tuomela, Raimo. 1989. Actions by Collectives. *Philosophical Perspectives* 3: 471–496.

UNEP. 2022. *Emissions Gap Report 2022: The Closing Window — Climate Crisis Calls for Rapid Transformation of Societies.* Nairobi: United Nations Environment Programme. www.unep.org/emissions-gap-report-2022.

Vanderheiden, Steven. 2008. *Atmospheric Justice: A Political Theory of Climate Change.* New York: Oxford University Press.

Whyte, Kyle Powys. 2017. Is It Colonial Déjà Vu? Indigenous Peoples and Climate Injustice. In *Humanities for the Environment: Integrating Knowledge, Forging New Constellations of Practice*, ed. Joni Adamson and Michael Davis, 88–104. London, New York: Routledge, earthscan.

Williams, Bernard. 1981. *Moral Luck: Philosophical Papers 1973—1980.* Cambridge: Cambridge University Press.

———. 1985. *Ethics and the Limits of Philosophy.* London: Fontana Paperbacks and William Collins.

Xu, Xiaoming, Prateek Sharma, Shijie Shu, et al. 2021. Global Greenhouse Gas Emissions from Animal-Based Foods Are Twice Those of Plant-Based Foods. *Nature Food* 2: 724–732.

Young, Iris Marion. 2011. *Responsibility for Justice.* New York: Oxford University Press.

Zheng, Robin. 2021. Moral Criticism and Structural Injustice. *Mind* 130 (518): 503–535.

CHAPTER 2

Calling All Collective Agents

All Hands on Deck

If climate change is to be mitigated to a level that human societies could still adapt to, organised collectives across the board need to make changes to how they operate. Although there have been some promising signs in recent years, such as ambitious new pledges by Japan and China,[1] the European Union hitting its 2020 emission reduction targets, and new climate legislation passed in the United States, the progress towards negotiating a sufficiently effective global treaty has been too slow to imagine that the situation will be resolved by top-level planning alone. Even within the existing measures, the reality does not match the rhetoric: almost 75% of pledges made under the Paris Agreement are insufficient to limit global warming to well below 2 °C, preferably 1.5 °C (Watson et al. 2019). There is no credible pathway to 1.5 °C currently in place (UNEP 2022). Instead, the world is on course for a 2.8 °C temperature rise by the end of the century, even with all the policies we currently have in place, which falls far short of the goals of the Paris Agreement (ibid.). But if we cannot reach the 1.5 °C target, or even the 2 °C target, every fraction of a degree still matters in limiting global warming and the harms it brings about (ibid.).

[1] In October 2020, Japan joined the European Union and New Zealand in pledging to go carbon neutral by 2050, while in the previous month China committed to do the same by 2060.

S. Hormio, *Taking Responsibility for Climate Change*,
https://doi.org/10.1007/978-3-031-51753-2_2

I will argue in this chapter that collective agents of all stripes are in a key position to innovate and deliver the meaningful structural-level changes required to address climate change. We need regulation, but we also need action across the board. Greenhouse gas (GHG) emissions from fossil fuels are embedded in our everyday practices of consumption and production, which I believe makes climate change too complex a problem to be resolved exclusively top-down. Hence, bottom-up problem-solving is required as well. With the scale and the imminence of the threat, all collective agents should make sure that their activities are as carbon neutral as possible. Just as importantly, collective agents can innovate better ways of working and new ways of doing things with lower emissions.

Large collective agents can arguably make a difference even in the absence of a binding international agreement, or often even despite what others do. This is because climate change is not a harm with an on-off level, but rather the harms and risk of harm keep increasing the more GHG we emit into the atmosphere. The more emissions, the more the global average temperature rises, leading to more climate change effects and harms. Carbon dioxide is the main driver of climate change, as it can accumulate for hundreds, even thousands of years.[2] The more cumulative atmospheric carbon there is, the more change there will be (Shue 2021, 7). Therefore, when a collective agent such as a state reduces its emissions, it helps to reduce the ultimate level or risk of harm (Farber 2008). Essentially, as climate change is not a monolithic harm with one threshold, climate harms and risk increase the more GHG emissions present in the atmosphere (Broome 2019). For these reasons, mitigation action by a state helps to reduce the ultimate level or risk of harm, regardless of what other countries do (Farber 2008). While no single agent, even a state or a large corporation, can mitigate climate change on its own, it could still impact how bad the overall effects will be.

A global climate treaty could be the optimal way to deal with climate change, at least in some sort of ideally rationalistic world. However, the idea that we need large governance units to resolve climate change has

[2] Water vapour contributes more to the overall greenhouse effect, but carbon dioxide controls the amount of water vapour in the atmosphere. This is because water vapour does not stay in the atmosphere for long periods of time, unlike carbon. When carbon dioxide concentrations rise, the average air temperatures go up, and the warmer the air, the more water vapour evaporates into the atmosphere. This amplifies the greenhouse effect, but carbon dioxide is the gas that sets the temperature. See www.earthobservatory.nasa.gov/features/CarbonCycle for more on this.

been challenged. Smaller units might work better to build the trust required to achieve workable emission reduction solutions (Ostrom 2010). Indeed, a polycentric approach to addressing climate change mitigation is steadily gaining prominence. Corporations, industries, non-governmental organisations, cities, and other such non-state actors have formed national and transnational networks to reduce their emissions (Andonova et al. 2009). For example, C40 is a network of megacities that supports them in collaborating effectively, sharing knowledge, and driving climate change action.[3] Another example is ICLEI—Local Governments for Sustainability, which is a global network of more than 2500 local and regional governments committed to sustainable urban development.[4] Polycentric actions like these can have a real impact on overall emissions (Ostrom 2010).

In any case, it is now widely recognised that we need an all-hands-on-deck approach to dealing with climate change. The United Nations Framework Convention on Climate Change (UNFCCC) has also seen a shift from a model of binding national emissions targets towards a more "catalytic and facilitative" model, which aims to actively involve non-state actors as well, such as cities, regional local governments, civil society actors, and businesses (Hale 2016). This usually takes the form of various voluntary targets. The UNFCCC meetings have been argued to play an important facilitative role in polycentric climate action, providing a place for different actors to meet (Lövbrand et al. 2017). Moreover, the UNFCCC Secretariat has taken on the role of a facilitator for multilateral and transnational action on climate change (Hale 2016). One example is their Race to Zero campaign, which seeks to rally non-state actors to take 'rigorous and immediate action' towards halving global emissions by 2030.[5] The increased interplay between state and non-state actors in international climate cooperation has been labelled 'hybrid multilateralism', characterised by deeply intertwined public and private authority, blurring the roles of state and non-state actors (Bäckstrand et al. 2017).

Some might argue that such calls for action misconstrue the concept of responsibility. If an agent fails to do their fair share of something, others are not necessarily responsible for taking on their burden, that is, for taking up the slack to achieve some collective good. The question is whether

[3] See www.c40.org.

[4] See iclei.org.

[5] See racetozero.unfccc.int/system/race-to-zero.

an agent's non-compliance can create moral duties for other agents. Some argue that justice requires each agent to contribute their fair share only (i.e., the burden of those who comply should not exceed their share under full compliance) and that any injustices that remain are the responsibility of the non-compliers (Miller 2011; Murphy 2000). Any additional duties generated by non-compliance would be unfair to those who comply. Others argue that such views are too conservative and fail to prioritise the possibly dire needs of the objects of the duty over the reasonable extra costs incurred by those who take up the slack (Stemplowska 2016). However, the costs associated with taking up the proverbial slack should not be overdemanding (Hohl and Roser 2011). The wider framework to which this debate applies is related to the way in which agents should act in non-ideal situations where not everyone is cooperative. Climate change is a global issue without an ideal system that would distribute burdens justly and that could deal with cases of non-compliance. If justice requires that each agent only does their fair share, that is, the share of the duty they would have under full compliance, then it seems that the whole starting point for discussing the responsibility of collective agents of all stripes is flawed. This book does not attempt to establish an overarching justice scheme for climate action. Still, the concept of taking up the slack is relevant in the wider sense of doing something that is not getting done otherwise: perceiving that there is an urgent need and being responsive to it, as long as it can be performed at a reasonable cost.

This chapter discusses *pro tanto* reasons, namely genuine reasons why we have to take action to tackle climate change now. While they might sometimes be outweighed in practice by other even weightier reasons once all things are taken into consideration, they are real reasons that cannot easily be sidelined. My premise is that given the scale and urgency of climate change, all collective agents have a *pro tanto* reason to reduce their emissions. This is both because of the harm that such emissions cause and the threat of what might happen in runaway climate change scenarios. Indeed, as emissions continue to rise, so does the potential for greater harm. We are too late to stop climate change now, as it is already underway, but we can still mitigate how dangerous and harmful it will become. Such a *pro tanto* reason is sufficient to ascribe a duty to reduce emissions beyond what any fair share scheme would look like (Hohl and Roser 2011). In other words, the fact that not all (collective) agents are doing what they should do if there were a fair global scheme is no reason for other agents not to do what they can. Note that I am not claiming that all agents have an *all things considered* reason to reduce their emissions. This is a further question that will depend on the circumstances of that agent. I will not attempt to generate a formula to determine the exact circumstances when *pro tanto* reasons turn into *all things considered* reasons, as that would be futile.

I begin with a general exposition of the responsibility of collective agents for climate change mitigation, suggesting that they should not only make their operations as carbon neutral as possible, but also play their part in achieving a carbon-neutral world by 2050, and cooperate with other actors as required to try to achieve systemic changes and support a pro-mitigation political environment.[6] After that I discuss why the individualistic framework has been getting so much attention, despite a focus on the responsibilities of collective agents being a much more fruitful framework when it comes to climate change. I then turn my attention to what Shue (2021, 130) describes as the "politico-economic web" of global vested and interlocking interests in the survival of the fossil fuel industry. This includes looking at the highly intertwined nature of many states and fossil fuel corporations, which I believe is at the root of the slow pace of tackling the problem at the international level.

The Responsibility of Collective Agents for Climate Change Mitigation

In this section, I will look at the responsibilities of organised collectives and formulate what the general structure of their responsibility could look like regarding climate change mitigation. I will make no attempt to present a detailed account of the responsibilities that any particular collective agents might have in relation to climate change mitigation. Such an account would first require an overall estimate of how mitigation costs should be fairly distributed among all non-state collective agents, either at the international or at the country level (or at the level of municipalities,

[6] I use 'agents' and 'actors' interchangeably in this book. Elsewhere, I have made a distinction between 'moral agents' and 'moral actors' (Hormio, forthcoming). I suggest that this helps to conceptualise why it makes sense to hold collective agents morally responsible for their actions: *even if* collective agents might not be moral agents, we can still hold them fully and holistically responsible. This is because as moral actors they are able to maintain and express moral positions of their own. "Moral actors" is a wider category than "moral agents", but the important point is that there is nothing that makes collective agents inherently incapable of being held fully responsible in moral matters.

districts, etc.). This is a job for participatory politics (see Chap. 6), not philosophy. One should then know what the contribution of any individual collective is, and given this contribution, what costs they should bear to address it. Differences in the operating field, size, available resources, and so on will also be reflected in the actual responsibilities of a collective agent. Although I will provide examples of some of these differences in the next chapter, a comprehensive, detailed account would be an impossible task. As I wrote in the introductory chapter, I am not focused on carving out the exact duties or obligations of any particular agent, as that is always contextual.

What I am after is the general structure of such responsibility. I am mainly interested in what agents have good reasons to do in light of climate change, and what they arguably should do from now on. The contours of the responsibility of collective agents will also help to delineate what their members should strive for (the topic of the next chapter). What suffices for my purpose is to note that with the scale and imminence of the threat, it would be hard to find a collective agent that would not have at least a *pro tanto* reason to make changes to the way it operates. If we are to reach the goal of mitigating climate change to a level that human societies could adapt to, organised collectives across the board need to make changes to how they operate. Furthermore, if a country's political elite is unwilling to move on mitigation, and government policies hinder rather than help mitigation action, it becomes even more crucial for other collective agents to take meaningful action. To begin the discussion, I stipulate the following responsibility:

> **The responsibility of collective agents regarding climate change mitigation**
> Collective agents should make their operations as carbon neutral as possible, playing their part in achieving a carbon-neutral world by 2050.

Going carbon neutral by 2050 is in line with the Paris Agreement and net zero strategies discussed in subsequent meetings and reports. The world going carbon neutral by that time means that carbon can still be emitted as long as there are enough carbon-negative actions to balance the emissions out.[7] The goal of net zero is to bring anthropogenic GHG

[7] Carbon neutrality allows for either carbon sinks or the removal of carbon from the atmosphere to offset (unavoidable) emissions.

emissions as close to zero as possible and to offset the remaining emissions by removing an equivalent amount of emissions from the atmosphere permanently (HLEG 2022). This can be applied at either a global or a local level, meaning that it can be used to address emissions from a single collective entity also. Importantly, the UN High Level Expert Group on Net Zero Commitments of Non-State Actors highlights that genuine net zero pledges should focus on concrete urgent steps across the value chain. They note that such pledges should not delay meaningful action and that buying carbon credits from voluntary markets cannot replace emission reductions. (Chapter 5 includes a discussion on the limits of offsetting.) The UN expert group has further clarified that net zero is incompatible with new investments in fossil fuel infrastructure and that any net zero pledges that involve such investments are greenwashing (HLEG 2022).

I believe that in the current system, which lacks a binding national and international plan, going carbon neutral by 2050 is a realistic, albeit ambitious demand for collective agents. The Paris Agreement (despite its limitations) can serve as a guide for collective agents when they formulate what they need to do to play their part in making their operations as carbon neutral as possible (and as soon as possible). Other documents, such as the annual UNEP Emissions Gap Report, can provide additional guidance. Due to the collective failure to act strongly and early enough, the cuts required are now deep and urgent, meaning that a system-wide transformation is required to reduce global emissions by 30–45% in order to get on track with the Paris Agreement goals (UNEP 2022). While these are not absolute benchmarks, they are clear indicators that urgent cuts in emissions are needed.

I will not discuss what 'as carbon neutral as possible' means in relation to feasibility constraints, as that would call for a chapter in its own right. There is no blueprint for determining exactly what a particular collective's plan should be in the absence of a national or international plan. It will depend not only on what their current emissions are and what they are doing about them but also on concerns about efficiency. Certain entities might not be the most efficient level to reduce emissions. Questions about the fair sharing of wider burdens also come into play. For example, while a collective entity delivering essential services (such as education or healthcare) in a developing country needs to be as efficient as possible, they arguably do not need to go carbon neutral, at least not as quickly as an entity delivering luxury holidays or other non-essential services. Still, 'as carbon neutral as possible' is meant to encompass the idea of as soon as possible in addition to as much as possible. Furthermore, depending on the field that the collective is in, or the kind of activities they engage in, their mitigation actions could focus on reducing other greenhouse gases, not just carbon dioxide. For example, certain actors in the food sector should aim to bring their methane emissions down, as meat production is a major source of methane (see Chap. 1). As the concept of carbon neutrality is often extended to include other greenhouse gases, measured in terms of their carbon dioxide equivalence, the formulation does not need to be amended on this account.

How such responsibility translates for any given organised collective depends on the collective agent and what it has already done (or omitted to do). It also depends on their core activities, namely their field of operation. While automobile industry actors should work towards increasing vehicle fuel efficiency, for example, those in the textile industry should look into ways of minimising waste and recycling materials. After all, textile production accounts for more emissions per year than international flights and maritime shipping combined (Ellen MacArthur Foundation 2017). In the case of universities, it could mean training future climate leaders and setting an example with ambitious strategies to reduce their emissions, activities that are increasingly taking place already. How emissions should be reduced most efficiently will also depend on the collective in question. Using the textile industry as an example again, this could include things like replacing virgin polyester with recycled polyester, phasing out polyester altogether in the manufacture of products, or designing clothes that are meant to last for years, contrary to disposable fashion.

Collective agents of all stripes should do what they can, including for-profit companies. However, I do not think that the responsibility of businesses to mitigate environmental harm is necessarily any greater than that of any other collective agents (except for the fossil fuel industry, which I will address below). Companies are not obliged to make their products completely carbon neutral in an environment that does not support this. If it raises the costs of their products to the point where they are at a serious competitive disadvantage and their survival is threatened, it is beyond the capability of the company to do so (Bowie 2013). This does not mean that all sectors or lines of business should somehow be safeguarded. Some should change their product range or services (see Chap. 4), while others should cease operations altogether (see Chap. 5).

The point is that it would be naïve and misleading to discuss the responsibilities of all for-profit entities under some one-size-fits-all plan, lumping small and medium-sized businesses together with large corporations, especially fossil fuel corporations. More generally, in considering what collective agents should do, we should keep in mind that they are embedded within a wider political and economic system and even the powerful among them are not omnipotent (Hormio 2017). This is why there is a need for not only regulation that changes the conditions in the system but also cooperation between different actors.

By way of illustration, the cultivation of palm oil is a major cause of tropical deforestation, which increases carbon emissions. Unilateral action by companies in sourcing so-called deforestation-free palm oil has not been very successful due to the complexity of the palm oil supply chain and the number of actors involved in it (Lyons-White and Knight 2018). For example, attempts to trace the origins of palm oil are complicated by the lack of direct demand-side engagement with the numerous suppliers, as they deal through intermediary trading companies. Multi-stakeholder action is required for the individual collective agents to implement the no-deforestation commitments made to their customers. More generally, cooperation within an industry is often required for any systemic change to be successful. Therefore, we should supplement the original formulation as follows:

> **The responsibility of collective agents regarding climate change mitigation***
> Collective agents should make their operations as carbon neutral as possible, playing their part in achieving a carbon-neutral world by 2050, and cooperate with other actors as required to try to achieve systemic changes.

At present, some collective agents are clearly in a position to make more of a difference than others when it comes to reducing their emissions. Organised collectives operating in industries that have heavy carbon footprints, such as energy, transport, construction, or agriculture, should pay particular attention to reducing their emissions as a matter of urgency. Collective agents operating in these sectors can have a major impact on emissions, even when they are constrained by the system they operate in. Heavy industry (including cement, steel, petrochemicals, glass and ceramics, and refining) is deemed one of the hardest areas to decarbonise, but doing so would reap enormous benefits. This is because heavy industry produces around 22% of global carbon emissions, out of which roughly

40% comes from the combustion of fossil fuels to produce industrial heat (Friedmann et al. 2019).

Take concrete, for example, widely used in construction. Concrete is made when water, sand, and gravel are mixed with cement. To make cement, natural limestone rock (calcium carbonate), which stores carbon naturally, is heated at very high temperatures for several days to turn it into quicklime (calcium oxide). Carbon dioxide is a by-product of this process, meaning that carbon that would otherwise have been stored in the rock for millions of years is released. The carbon released accounts for a little over half of cement's total GHG footprint, while the rest comes from fossil fuels used to heat cement kilns. Cement kilns require a lot of energy at very high temperatures: around 1100–1400 °C. Currently, this industrial heat is produced by fossil fuel combustion. One high-impact way to reduce emissions by heavy industry actors would be to produce industrial heat with renewables. A recent breakthrough in the field of concentrated solar thermal technology means that renewables can produce industrial heat at very high temperatures and for the timespan required. By utilising a field of mirrors that acts as a magnifying glass for sunbeams, an ultra-high concentration solar system is capable of achieving temperatures up to 1500 °C.[8] This is an important development, although the technology still needs to be scaled up. Another way to significantly reduce emissions from concrete is carbon sequestration, namely long-term storage of carbon. There is much that industry actors can already do in this field. Promising carbon sequestration technologies include making concrete from carbonate rocks, produced from carbon dioxide captured from power plants, making the concrete carbon neutral.[9]

Companies operating in heavy industry will need to invest in new technologies to be able to switch over to using carbon sequestration and ultra-high concentrated solar. This can incur considerable additional costs for them as the existing fossil fuel technology is replaced. However, these investments will pay off in the long run. First of all, the energy source used is itself free: no additional drilling or digging is required once concentrated solar is in use. Second of all, the world needs to turn carbon neutral, so future policies will increasingly penalise those actors relying on old

[8] Heliogen (www.heliogen.com) achieves industrial heat temperatures by getting thousands of computer-controlled mirrors to reflect sunlight onto one spot, utilising artificial intelligence to train the mirrors.

[9] Blue Planet (www.blueplanet-ltd.com) produces carbonate rocks in this way.

technologies with heavy emissions. However, the speed at which the policy changes will take place is not set in stone and will depend in part on how willing the big industry actors are to accelerate the process of going carbon neutral (or carbon negative with sequestration). Market forces will have their say as well, but part of this process is decided internally: are we among the first to invest in the new technologies, or will we bide our time and follow others only when we must? There are potential benefits and drawbacks to both approaches: wait, and the new green technology will become cheaper; wait too long, and you fall behind your competitors. In the next chapter, I will discuss the role of individual members in pushing collective entities to take climate action.

Reducing their own emissions is not the only way in which collective agents can make an impact. They can also affect the political environment. Some collective actors lobby for increased regulation in their field, as effective regulation will make it easier for them to become proactive in mitigation. Ambitious sector-specific standards, combined with flexible instruments that leave room for voluntary private-sector initiatives, are among policy measures that can facilitate corporate climate leadership (Dupuis and Schweizer 2019). A regulative framework that sets shared standards for competitors will be more crucial in some fields, like aviation, than in others. Governments and politicians usually want strong political support before they implement large-scale changes. Active engagement in emissions reduction by various collective agents can send a strong signal to decision-makers. I have in mind things like pension funds decarbonising their investments, companies installing solar panels on their roofs, or industry bodies issuing statements in support of mitigation action. Once clear political support is in place, we can begin to see the desirability of macro-level solutions, such as changes in subsidies for fossil fuels. Therefore, the formulation should be made a little more demanding:

> **The responsibility of collective agents regarding climate change mitigation****
> Collective agents should make their operations as carbon neutral as possible, playing their part in achieving a carbon-neutral world by 2050, and cooperate with other actors as required to try to achieve systemic changes and support a pro-mitigation political environment.

I have argued that collective agents have a *pro tanto* reason to make their operations as carbon neutral as possible, both due to the urgency and the scale of the threat, as well as the lack of sufficient top-down action. How organised collectives act is decided for the most part by their members, but always within the wider social setting in which the collective is embedded. That is why even when discussing what collective agents should do, we must not forget the importance of the wider structures that they are part of (Hormio 2017). However, it is important to keep in mind that societal structures are not somehow inevitable, but built by people. Therefore, they can also be changed by people. After all, humans are nothing if not inventive in coming up with new ways of organising their lives together (Graeber and Wengrow 2021).

Human inventiveness also comes to the fore with new technology, like the ultra-high concentrated solar already mentioned. New inventions affect the available options for collective agents. In the face of new technology and other innovations, collective agents must choose how to react. I discuss new lower-carbon alternatives to meat and dairy products in Chap. 4. Many meat industry actors have not stood idly by as new innovations emerge and gain ground with consumers. Instead, they have been fighting hard to slow down their popularity, including lobbying to get government officials to ban words like 'butter' and 'milk' on non-dairy products, or trying to mislead the public on the qualities of the vegetarian alternatives (Monbiot 2023). For example, Meat & Livestock Australia's online platform Good Meat describes plant-based meat alternatives as "ultra-processed manufactured food", which compare poorly with "natural, nutritious and sustainable Australian red meat".[10] The website has been created on behalf of the country's red meat and livestock industry. On the subject of climate change, it vaguely states thus: "Although foods vary in greenhouse gas emissions, it's also important to look at the bigger picture – the nutritional value of foods as well as the greenhouse gas emissions", going on to compare lettuce leaves to beef. It also misleadingly attributes the biodiversity loss caused by soybean farming to "plant-based fake meats", as soy is the main raw material for these products, conveniently ignoring that the vast majority of soy in the world goes to feed livestock for meat and dairy producers.

Engaging in such measures is, of course, not the only path available. Instead, it is up to the collective agent how they respond: if they see change as a threat only, or also as an opportunity. For example, Makuliha, a family-run business specialising in the development of different flavoured

[10] See www.goodmeat.com.au/health-nutrition/fake-meat.

cold cuts and sausages made from Finnish meat, has recently branched out to include vegetarian cold cuts and sausages.[11] Smaller family-run businesses are not the only ones that can adapt to the times: another example is Pouttu, a large meat company in Finland. When they launched their vegetarian range in 2019, aimed at offering plant-based products to those who normally eat meat, they expected the products to amount to 5% of sales, but within a year, this figure was already 10–15% (Vähäsarja 2020). Such adaptable actors can face pressure from within the industry: Atria, the leading meat company in Finland, made a complaint to officials about Pouttu describing its products as vegetarian meat, with officials forcing the company to change their packaging (Paananen 2019). This is in line with the lobbying efforts of other animal industry actors to ban vegetarian products from using words like ‘butter’ and ‘milk’ on their labels. (Monbiot 2023 wryly observes that their stand on peanut butter and coconut milk is anyone’s guess.) The point is that collective agents choose how they react to things, including climate change and the pressure to change the status quo that it brings with it. After all, collective agents are not pre-destined machines, but entities that make decisions and react to the world around them.

Taking the discussion back to the broader level, the remit we give to collective agents and what we expect of them in return is up for debate. Values in society are constantly being strengthened or re-negotiated. What has been noticeable in recent years is the trend of demanding more from companies, not just from grassroots social movements, but also from the inside. One example of values changing is how in 2019 the chief executive officers (CEOs) of Business Roundtable, a lobby association representing the CEOs of America’s largest corporations, overturned a 22-year-old policy statement that described a corporation’s principal purpose as maximising shareholder return (à la Milton Friedman). Instead, the new Statement on the Purpose of a Corporation recognises that in addition to generating long-term value for shareholders, companies should also deliver value to their customers, invest in their employees, deal fairly and ethically with their suppliers, and support the communities in which they work (Business Roundtable 2019). Much of the pressure for change comes not just from the wider society, but also from the members of collective agents, be they employees or shareholders.

[11] The vegetarian range produced by www.makuliha.fi utilises pea protein.

In the next chapter, I consider what members of collective agents should do to push for collective climate action, as I believe that our responsibility *qua* members of organised collectives can often be our primary responsibility in relation to climate change.[12] But before that, for the rest of this chapter, I will turn my attention to the fossil fuel industry, which is brimming with collective agents with extremely heavy carbon footprints and excessive influence on climate policy. They are a special case when it comes to climate change responsibility and should be discussed separately from the general rubric of responsibility of collective agents.

Misleading Framing and Evasion of Responsibility

In this section, I will outline why the fossil fuel industry presents a special case in terms of the climate responsibility of collective agents. This is due to not only their large carbon footprints but also heavy lobbying and misdirecting the public debate.

While most large collective agents will almost inevitably need to be involved in order for mitigation efforts to be adequate given the urgency and scale of the climate problem, much ink has been spilled over what individuals should do. Some of this is to be expected. When faced with a problem or threat, it is a natural human impulse to want to do something about it. It is a human need to seek to exercise some control over our surroundings. Media stories about ways to reduce one's individual carbon footprint respond to this need. Yet the framing which focuses on individual responsibility for climate change has not always emerged naturally. Instead, the somewhat misleading framing has been partly manufactured to obfuscate responsibility. More specifically, corporations with vested fossil fuel interests have deliberately used this framing device to deflect attention away from what *they* should be doing.

Not all of the remaining fossil fuels in the ground can be used, as doing so would spell certain climate chaos. Despite this, the fossil majors have been and still are busy campaigning against regulation. Climate science emerged in the 1950s and was discussed in the US Congress in 1965. In response, the fossil fuel industry began a sustained lobbying effort against regulation (Oreskes and Conway 2010). The consensus across parties in

[12] The primary responsibility depends on the person, including their exact circumstances and personal qualities. Not everyone is cut out to be an activist or a leader, but nearly everyone is a member of a collective of some sort where they could push for change.

the United States on the importance of mitigation action evaporated, turning climate change into the divisive issue that it is today between Democrats and conservative Republicans. In the coming decades, as the science became more robust and the urgency of the threat clearer, instead of heeding the advice of its own scientists who affirmed the anthropogenic nature of the changes, Big Oil continued to ramp up their efforts at misleading the public about climate science (Oreskes and Conway 2010). Even after the Paris Agreement, it is estimated that nearly US$200 million is spent each year on lobbying to block, delay, or control climate change policy, with Chevron, BP, and ExxonMobil leading these activities (InfluenceMap 2019). Research suggests that exposure to misinformation about climate change makes individuals less likely to want to take action to reduce emissions (Jolley and Douglas 2014).

While earlier fossil fuel industry communications over-emphasised scientific uncertainties to delay meaningful climate change policies, Geoffrey Supran and Naomi Oreskes (2021) show in their computational textual analysis how a subtler rhetoric is used today to protect fossil fuel interests. The focus of their research article is on how ExxonMobil has purposefully framed its communications to individualise climate change responsibility in recent decades. This includes presenting consumer demand as the driving force for fossil fuel production and GHG emissions, underplaying the role of corporations who supply the oil, gas, and coal. The growing global demand for energy is presented as an inevitable fact, and it is implied that fossil fuels are essential to meet societies' needs. In the narrative thus formed, fossil fuel firms are only passively responding to this demand. Supran and Oreskes (ibid.) also observe that ExxonMobil's advertorials focus on the need for individual actions to mitigate climate change, such as becoming more energy efficient. Conserving energy is presented as the solution, not addressing the supply side, or constraining the extraction of fossil fuels. In other words, climate change becomes a downstream problem caused by consumer demand, to be resolved by energy efficiency, whereas the nature of climate change as an upstream problem caused by the supply side is swept under the carpet. Notoriously, fossil fuels were not even mentioned in the Paris Agreement in 2015.

This strategy, which replaces concerns about corporate responsibility with talk of individual responsibility to conserve energy, has also been utilised by other fossil fuel majors. For example, the concept of 'carbon footprint', used to highlight the emissions of an individual, was coined not by some environmental charity, but by a PR company (Brulle and

Werthman 2021). The notion was popularised by BP as part of its US media campaign in 2004–2006 (Supran and Oreskes 2021). Campaigns like these frame individuals as consumers first and foremost. The idea that concerned citizens can come together to demand policy changes gets pushed aside as the focus is on what you choose to do as a consumer.

Misleading individualistic framing is only one strategy used by the fossil fuel industry and carbon majors; greenwashing is another. Greenwashing campaigns by PR companies attempt to give the impression that the client organisation is a socially responsible and trustworthy actor (Brulle and Werthman 2021). For fossil fuel agents, the idea is to present marketing campaigns that create the impression that the corporation is at the forefront of the energy revolution, rapidly transitioning to low-carbon energy through measures such as carbon capture and storage, even when they represent only a fraction of their activities. As the environmental law charity ClientEarth points out, while there may be nothing wrong with companies highlighting their green work in their advertisements, this becomes a problem when a misleading impression is created of the overall business model and its environmental impact.[13] Misleading communications do not necessarily have to contain any factual errors but can be a matter of the context in which the facts are placed, or the absence of a proper context. Greenwashing is consequently about misleading your customers and the public about the green credentials of your products.

As outright climate denialism is no longer a feasible strategy due to the strength of the scientific evidence, framing and greenwashing tactics have become the new norm in the fight against climate action. For these activities, the industry has turned to top PR and advertising agencies, which can be viewed as enablers of the fossil fuel industry, along with lobbying organisations, think tanks, and law firms (Mufson 2022). PR firms have developed campaigns for fossil fuel corporations that frequently rely on third-party groups to engage with the public and criticise opponents (Brulle and Werthman 2021). By changing the context in which climate communications take place, PR firms have helped fossil fuel companies engage in "climate obstruction", which has been described as an "even more insidious" tactic than the old denialist one (Aronczyk quoted in Joselow 2021). Some consulting firms have also engaged in misleading communications on behalf of oil and gas companies by utilising techniques such as designing and running campaigns that masquerade as grassroots

[13] See www.clientearth.org/projects/the-greenwashing-files.

activism (Tabuchi 2020). Engaging in such activities can lead to additional responsibilities for the collective agents involved, in terms of both mitigation and compensation. Interestingly, several court cases have already been brought over the inaccuracy of green claims and commitments (Setzer and Higham 2023). As consumer litigation against ESG (environmental, social, and governance) claims on product packaging and advertisements is on the rise (Clifford Chance 2022), this could be seen as part of a wider trend in litigation over false advertisements and deceptive trade practices related to ESG investments.

Naturally, not all collective agents with heavy carbon footprints have engaged in disinformation, and there are significant differences in terms of the amount of GHG emissions produced by different actors. These aspects should play their part in responsibility judgments. The polluter pays principle is widely accepted and is adopted as part of the legislation in the European Union and the United States, for example. According to the principle, an agent should be accountable for the environmental damage they have caused. For example, a factory that is polluting a river as a side-effect of manufacturing its products should bear the costs of preventing such damage in the future and compensate for the harm already caused. 'Polluter pays' can also be attached to prior blameless action, where pollution was both unintended and unforeseen. It is based on causality and entails liability, but the principle remains neutral with regard to culpability. Applying 'polluter pays' to climate change would seem to imply that those entities responsible for fossil fuel emissions should pay for the damage. Thus, it could be argued that the 90 entities to which 63% of worldwide industrial CO_2 and methane emissions from 1751 to 2010 can be attributed (Heede 2014) should pay for mitigation and adaptation, perhaps also compensation. However, applying it straightforwardly over such a long timeframe—hundreds of years into the past—can run into difficulties.

Looking at more recent decades, just 20 companies have contributed 35% of all energy-related carbon dioxide and methane emissions since 1965 (Heede 2019). Saudi Aramco (the Saudi Arabian Oil Company), Chevron, Gazprom, ExxonMobil, National Iranian Oil, BP (British Petroleum), Royal Dutch Shell, Coal India, Pemex (Petróleos Mexicanos) and Petróleos de Venezuela comprise the top 10 polluters from 1965 to 2017. Responsibility for historical emissions has been opposed due to lack of knowledge about climate change at the time of the emissions and the unfairness of penalising current generations for the pollution caused by their predecessors (e.g., Caney 2009). It is important to note that such objections do not apply here. Firstly, the emissions under consideration have emerged since climate science was known to the fossil fuel corporations. Secondly, the question of predecessors is different for companies with role-occupiers, rather than previous generations of citizens. Responsibility for past wrongdoings seems to transfer more easily in the corporate setting, due to the lineated roles, corporate strategies, and shareholder concerns. That said, there is still the complication of which emissions should count on the ledger of a particular agent. Production- and consumption-based emissions accounting offer different ways of calculating which emissions belong to which agent's GHG record. The ownership and historical roots of some of these companies create further complications, as I will argue in the next section, where I turn to issues relating to the responsibility of these major polluters.

National and Private Interests

The world needs to decarbonise its energy infrastructure rapidly. Reliance on fossil fuels must be halted, and quickly. Fossil fuel assets need to decline swiftly because they work against the clean energy transition and lock in GHG emissions for decades to come (UNEP 2022, 39). The most effective way to tackle climate change would be to strictly regulate the fossil fuel industry through binding international regulations and domestic policies. However, as has already been noted, such regulation has proceeded slowly. There are many reasons for this, the most obvious being that the fossil fuel infrastructure is deeply embedded in modern societies and infrastructures cannot be changed overnight. The fossil fuel industry has also been a very active lobbyist and disseminator of disinformation, as discussed in the previous section. Not only this, but fossil fuel interests are widely dispersed within many countries, with different actors benefitting from the continuing exploitation of fossil fuels. Matto Mildenberger (2020) suggests that in order to understand how the push-and-pull in climate policies has been shaped in different countries, we need to bear in mind that the interests of carbon polluters have been advanced not only by the political right via industrial business associations, but also by the left through industrial unions. The latter fear the job losses that a shift away from fossil fuels could cause for those working for dirty energy utility companies and other carbon-dependent sectors. Mildenberger (ibid.) calls this the 'double representation' of carbon polluters and argues that it has ensured that the interests of the carbon majors have been accommodated in policymaking regardless of which party is in power.[14] The stronger the double representation within a country, the less likely it is that ambitious domestic climate policies will emerge.

While fossil fuel companies are powerful actors in and of themselves, they also have powerful enablers. Shue (2021, 128) highlights that government allies are not the only ones promoting fossil fuel interests through their subsidies and lax regulation, but that financial institutions, such as banks and hedge funds, also have a significant role to play in enabling additional production of fossil fuels. By lending money to fossil fuel corporations, or through investing in them, these financial institutions allow the corporations to engage in activities that expand fossil fuel production, such as fracking and building new pipelines. One example is JP Morgan Chase, which has financed fossil fuel companies to the tune of $434 billion since the Paris Agreement in 2015 (Merleaux et al. 2023). Although leading the pack, it is sadly far from alone in engaging in these harmful practices: the 60 biggest banks in the world have financed fossil fuels by $5.5 trillion in the seven years between 2016 and 2022 alone (ibid.). Some banks have started to adopt policies against such destructive practices, but they are still in the minority.

States and their governments are responsible for regulating the fossil fuel corporations. I will discuss the responsibility of members of publicly listed corporations in the next chapter. However, it is important to underline how often state and corporate interests are intertwined when it comes to fossil fuels. In particular, I wish to highlight that the embeddedness of national policies with fossil fuel interests also runs deep for the simple reason that many of the carbon majors are owned or controlled by states. Out of the top ten polluters listed in the previous section, Chevron, ExxonMobil, BP, and Royal Dutch Shell are publicly listed companies, while Saudi Aramco, Gazprom, National Iranian Oil, Coal India (Coal India Limited), Pemex, and Petróleos de Venezuela are state-owned or state-controlled entities. In a national oil company (NOC), the government controls the company, usually through special ministers appointed to oversee its operations. Some of the NOCs also allow for investors: while the majority of Gazprom is owned by the Russian government, its shares

[14] Coal mines have been ascribed with the ability to create communities, through shaping towns over decades, their history of strong unions, and perhaps even through the collective nature of mining itself (Meyer 2020). This effect can be seen even after the coal mines have been shut, in predicting local voting behaviour, for example.

are also traded on stock markets. The same formula of majority ownership by government, with publicly listed shares, also applies to Coal India and Pemex, albeit with different ratios of ownership. The latest entrant was Saudi Aramco, which was listed on the Saudi Stock Exchange at the end of 2019. The state-owned entities vary in terms of who directly benefits from them. For example, Saudi Aramco is controlled and owned by the House of Saud, the ruling royal family, whereas Norway's Equinor pays billions of dollars into the state's coffers in dividends each year.

The state has the power at NOCs. In international climate talks, the position of states with large oil assets is aligned with that of their national oil company. According to Bryony Worthington, executive director of the European chapter of the Environmental Defense Fund, the interests between nations and NOCs are inseparable: "There is no distance between NOCs and their governments in negotiations – they are the same" (Harvey 2019). The close link between oil and the budget of an NOC state is acknowledged also by the oil-rich nations. For example, Iran's former president, Hassan Rouhani, has stated that "Budget and oil are twins", noting how separating the two will be difficult as Iran prepares to reduce its reliance on oil revenues (Tehran Times 2019). These blurred lines raised eyebrows recently when the United Arab Emirates gave several of their NOC employees roles in the office hosting the UN climate summit in the UAE in 2023 (Stockton and Carter 2023).

NOCs produce most of the world's oil but, as noted by Shue (2021, 122), this "fact about state-ownership is often overlooked by climate activists". Perhaps part of the reason for this is that many of these oil companies operate in non-democratic, or only partially democratic states, where activism is dangerous. The same applies to other nationally owned fossil fuel companies, such as gas producers. Turkmenistan is currently responsible for the highest number of large methane leak events, where potent GHG leaks into the atmosphere, likely from old and poorly maintained equipment used in natural gas production (Carrington 2023). Despite this, it would not make much sense to call for climate activism from citizens living inside the repressive and secretive dictatorship. Relatedly, what we can reasonably expect from citizens in terms of knowledge about political issues depends on the availability and quality of information on a given topic. What could be labelled as the 'requirement to know' is usually thought to be lower for citizens in authoritarian regimes than in democracies. David Miller (2007, 130) notes about the epistemic situation of citizens in autocratic nations that "[j]ust as we cannot expect people to make superhuman efforts to oppose a regime, so we cannot expect them to stand firm against the propaganda barrage that descends on them". Conversely, "the more open and democratic a political community is, the more justified we are in holding its members responsible for the decisions they make and the policies they follow" (ibid., 130).

Applying this logic, we should expect more in terms of knowledge about the urgency of climate change mitigation efforts from citizens of democratic states with NOCs, like Norway, than from the average citizen of Turkmenistan or Saudi Arabia. Additionally, arguably more activism is called for within democratic countries that purchase and consume the fossil fuels produced by non-democratic countries. Leif Weinar (2018, 3) describes wars fought over oil as disasters that Western consumers have paid for. Climate change follows the same logic. The twist in the tale is the efficient and relentless disinformation tactics employed by the oil industry, as described in the previous section, which makes it somewhat unclear what we can reasonably expect from an average person in terms of climate knowledge, at least looking back a decade or two. Be that as it may, I contend that what should be highlighted more in public debates and activism inside democratic states is how governments seem happy to conduct business with non-democratic NOCs and what this means in terms of climate change. This is of course something that has been questioned increasingly due to the Russia-Ukraine war, but it seems that there is still room for more activism against the acceptability of certain types of transactions.

When looking at historical emissions, the distinction between NOCs and the investor-owned companies becomes more blurred. In Iran, the oil industry started in the early 1900s under the moniker Anglo-Persian Oil Company. The industry went through a period under a holding company called Iranian Oil Participants Ltd in the middle of the century, with founding members including earlier incarnations of BP and Chevron, along with Royal Dutch Shell, before the oil assets were nationalised and National Iranian Oil was born in 1948. Saudi Aramco started out as the California-Arabian Standard Oil Company in 1933, founded by Chevron (itself known as the Standard Oil Co. (California) at the time). In 1944, the name was changed to Arabian American Oil Company (Aramco) and ownership of the company diversified, with owners including earlier components of what is now ExxonMobil. The company gradually reduced the ownership of foreign companies and completed its nationalisation in 1980. In a similar fashion, privately owned oil companies used to operate in Venezuela before the oil industry was nationalised in the 1970s and

Petróleos de Venezuela created.[15] I believe that this history adds yet another layer to why historical emissions are such a sensitive subject for industrial states in climate negotiations.

Another aspect to the global debate on responsibility for climate change action is the uneven power dynamics between poorer and richer nations. Although the parties to the UNFCCC already agreed in 1992 that because industrialised countries have produced most of the GHG emissions, they should also take the lead in climate action, questions of climate equity have been at the centre of climate negotiations ever since. There is a body of literature on who should pay for climate action, and I will not go into the burden-distribution problem in this book. What I wish to note here is that what NOCs decide to do in poorer countries is partially up to what the rich nations do or omit to do. Ecuador proposed in 2007 that it would leave around one billion barrels of its oil reserves in the ground, as long as other countries paid into a fund to compensate Ecuador for half of the oil's estimated value (Einhorn and Andreoni 2023). The plan was proposed by the country's minister of energy at the time, Alberto Acosta. In addition to supporting climate mitigation, the plan would have also protected the Indigenous people living in the Yasuní National Park area and biodiversity in the Amazon rainforest. However, despite a promising start, including the UN agreeing to manage the fund, contributions were meagre.[16] The country's leaders gave Petroecuador the green light to drill for oil, arguing that they could not afford to keep it in the ground as up to one in four Ecuadorian children were suffering from malnutrition (Einhorn and Andreoni 2023).[17] Interestingly, the Ecuadorian people voted in August 2023 to end drilling in the Amazon, meaning that Petroecuador will have to shut down the oil field in the Yasuní National Park area (Glatsky et al. 2023).[18]

An additional threat is posed by the fact that the clock is ticking for fossil fuels. Ecuador's president, Guillermo Lasso, stated in 2022 that, "Now that the global trend is to abandon fossil fuels, the time has come to extract

[15] Pemex nationalised as early as 1938.

[16] Partly due to concerns over the populist leader at the time, Rafael Correa (Einhorn and Andreoni 2023).

[17] It is doubtful whether the communities near the oil development sites in the Ecuadorian Amazon would stand to benefit, with the Confederation of Indigenous Nationalities of Ecuador leading protests against oil (Einhorn and Andreoni 2023).

[18] As this example indicates, citizens can and should play a bigger role in climate policies, something I will discuss in Chap. 6.

every last drop of benefit from our oil, so that it can serve the poorest while respecting the environment", while the Democratic Republic of Congo is auctioning oil blocks within rainforests (Einhorn and Andreoni 2023). This is an example of "final harvest" thinking, which poses a serious threat to reducing emissions quickly enough (Shue 2021, 124), and which is a logic also espoused by corporations in rich countries, for example, through fracking. New pipelines and other oil and gas projects have been recently approved in the United States and elsewhere, with new fossil fuel extraction projects running into hundreds in their numbers in the past year alone (Bearak 2023). Without structural changes, the impact of our continuing reliance on fossil fuels spells further doom. It is a highly harmful web that needs dismantling from many directions.

Concluding Remarks

The starting point for the chapter was that we need all hands on deck to solve the climate crisis. Collective agents of all stripes can innovate and deliver meaningful structural-level changes to mitigate climate harms. I have suggested that collective agents should make their operations as carbon neutral as possible, as well as play their part in achieving a carbon-neutral world by 2050, cooperating with other agents as required to try to achieve systemic change. The fossil fuel industry presents a special case for climate responsibility. Not only can the majority of emissions be traced back to them, but the fossil fuel majors have pushed false narratives and misleading claims about climate change into the public debate. The intertwined nature of many states and fossil fuel corporations has slowed down mitigation action.

References

Andonova, Liliana B., Michele M. Betsill, and Harriet Bulkeley. 2009. Transnational Climate Governance. *Global Environmental Politics* 9 (2): 52–73.

Bearak, Max. 2023. It's Not Just Willow: Oil and Gas Projects Are Back in a Big Way. *The New York Times*, April 6.

Bowie, Norman E. 2013. Morality, Money, and Motor Cars Revisited. In *Business Ethics in the 21st Century. Issues in Business Ethics*, ed. Norman E. Bowie, vol. 39, 131–146. Dordrecht: Springer.

Broome, John. 2019. Against Denialism. *The Monist* 102 (1): 110–129.

Brulle, Robert J., and Carter Werthman. 2021. The Role of Public Relations Firms in Climate Change Politics. *Climatic Change* 169: 8.
Business Roundtable. 2019. *Statement on the Purpose of a Corporation*. https://opportunity.businessroundtable.org/ourcommitment.
Bäckstrand, Karin, Jonathan W. Kuyper, Björn-Ola Linnér, and Eva Lövbrand. 2017. Non-state Actors in Global Climate Governance: From Copenhagen to Paris and Beyond. *Environmental Politics* 26 (4): 561–579.
Caney, Simon. 2009. Justice and the Distribution of Greenhouse Gas Emissions. *Journal of Global Ethics* 5 (2): 125–146.
Carrington, Damian. 2023. Revealed: 1,000 Super-Emitting Methane Leaks Risk Triggering Climate Tipping Points. *The Guardian*, March 6.
Clifford Chance. 2022. A Different Shade of Greenwashing: US Litigation Over Advertising and Product Packaging Claims Accelerates. *Clifford Chance Business & Human Rights Insights*, August 17. www.cliffordchance.com/insights/resources/blogs/business-and-human-rights-insights/2022/08/a-different-shade-of-greenwashing-us-litigation-over-advertising-and-product-packaging-claims-accelerates.html.
Dupuis, Johann, and Remi Schweizer. 2019. Climate Pushers or Symbolic Leaders? The Limits to Corporate Climate Leadership by Food Retailers. *Environmental Politics* 28 (1): 64–86.
Einhorn, Catrin, and Manuela Andreoni. 2023. Ecuador Tried to Curb Drilling and Protect the Amazon. The Opposite Happened. *The New York Times*, January 14.
Ellen MacArthur Foundation. 2017. *A New Textiles Economy: Redesigning Fashion's Future*. Ellen MacArthur Foundation.
Farber, Daniel A. 2008. The Case for Climate Compensation: Justice for Climate Change Victims in a Complex World. *Utah Law Review*: 377–413.
Friedmann, S. Julio, Zhiyuan Fan, and Ke Tang. 2019. *Low-Carbon Heat Solutions for Heavy Industry: Sources, Options, and Costs Today*. New York: Columbia University Center on Global Energy Policy.
Glatsky, Genevieve, José María León Cabrera, and Thalíe Ponce. 2023. Establishment Leftist and Newcomer Businessman Appear Headed to Ecuador Runoff. *The New York Times*, August 20.
Graeber, David, and David Wengrow. 2021. *The Dawn of Everything: A New History of Humanity*. London: Allen Lane.
Hale, Thomas. 2016. "All Hands on Deck": The Paris Agreement and Nonstate Climate Action. *Global Environmental Politics* 16 (3): 12–22.
Harvey, Fiona. 2019. Secretive National Oil Companies Hold Our Climate in Their Hands. *The Guardian*, October 9.
Heede, Richard. 2014. Tracing Anthropogenic Carbon Dioxide and Methane Emissions to Fossil Fuel and Cement Producers, 1854–2010. *Climatic Change* 122: 229–241.

———. 2019. *Carbon Majors: Update of Top Twenty companies 1965-2017*. Press release, October 9. Climate Accountability Institute, Colorado. https://climateaccountability.org/wp-content/uploads/2020/12/CAI-PressRelease-Top20-Oct19.pdf.

Hohl, Sabine, and Dominic Roser. 2011. Stepping in for the Polluters? Climate Justice Under Partial Compliance. *Analyse und Kritik* 33: 477–500.

Hormio, Säde. 2017. Can Corporations Have (Moral) Responsibility Regarding Climate Change Mitigation? *Ethics, Policy & Environment* 20 (3): 314–332.

———. forthcoming. Collective Agents as Moral Actors. In *Collective Responsibility: Perspectives on Political Philosophy from Social Ontology*, Studies in the Philosophy of Sociality, ed. Säde Hormio and Bill Wringe. Springer.

HLEG. 2022. *Integrity Matters: Net Zero Commitments by Businesses, Financial Institutions, Cities and Regions*. United Nations' High-Level Expert Group on the Net Zero Emissions Commitments of Non-State Entities.

InfluenceMap. 2019. *Big Oil's Real Agenda on Climate Change: How the oil majors have spent $1bn since Paris on narrative capture and lobbying on climate*. An InfluenceMap Report, March 2019. https://influencemap.org/report/How-Big-Oil-Continues-to-Oppose-the-Paris-Agreement-38212275958aa21196dae3b76220bddc.

Jolley, Daniel, and Karen M. Douglas. 2014. The Social Consequences of Conspiracism: Exposure to Conspiracy Theories Decreases Intentions to Engage in politics and to Reduce One's Carbon Footprint. *British Journal of Psychology* 105: 35–56.

Joselow, Maxine. 2021. How Big Oil Relies on Some PR Firms to Block Climate Action. *The Climate 202, The Washington Post*, November 30.

Merleaux, April, Caleb Schwartz, Ruth Breech, et al. 2023. *Banking on Climate Chaos: Fossil Fuel Finance Report 2023*. Rainforest Action Network et al. www.bankingonclimatechaos.org.

Lyons-White, Joss, and Andrew T. Knight. 2018. Palm Oil Supply Chain Complexity Impedes Implementation of Corporate No-Deforestation Commitments. *Global Environmental Change* 50: 303–313.

Lövbrand, Eva, Mattias Hjerpe, and Björn-Ola Linné. 2017. Making Climate Governance Global: How UN Climate Summitry Comes to Matter in a Complex Climate Regime. *Environmental Politics* 26 (4): 580–599.

Meyer, Robinson. 2020. The Secret Political Power of Fossil Fuels. *The Weekly Planet, The Atlantic*, October 27.

Mildenberger, Matto. 2020. *Carbon Captured: How Business and Labor Control Climate Politics*. Cambridge, MA: The MIT Press.

Miller, David. 2007. *National Responsibility and Global Justice*. New York: Oxford University Press.

———. 2011. “Taking Up the Slack? Responsibility and Justice in Situations of Partial Compliance.” In *Responsibility and Distributive Justice*, edited Carl Knight and Zofia Stemplowska, 230–245. New York: Oxford University Press.
Monbiot, George. 2023. ‘Let Them Eat Lentils’ Won’t Save Us from Animal Farming – We Must Embrace Meat Substitutes. *The Guardian*, February 1.
Mufson, Steven. 2022. More Than 450 Scientists Call on PR and Ad Firms to Cut Their Ties with Fossil Fuel Clients. *The Washington Post*, January 19.
Murphy, Liam. 2000. *Moral Demands in Nonideal Theory*. New York: Oxford University Press.
Oreskes, Naomi, and Erik M. Conway. 2010. *Merchants of Doubt: How a Handful of Scientists Obscured the Truth on Issues from Tobacco Smoke to Global Warming*. New York: Bloomsbury Press.
Ostrom, Elinor. 2010. Polycentric Systems for Coping with Collective Action and Global Environmental Change. *Global Environmental Change* 20: 550–557.
Paananen, Karoliina. 2019. Asiakirjat paljastavat Atrian kannelleen Poutun kasvilihasta – Pouttu yhä uhmakkaana: “Sääntö-Suomea haastettava”. *Suomen Kuvalehti*, December 11.
Setzer, Joana, and Catherine Higham. 2023. *Global Trends in Climate Change Litigation: 2023 Snapshot*. London: Grantham Research Institute on Climate Change and the Environment and Centre for Climate Change Economics and Policy, London School of Economics and Political Science.
Shue, Henry. 2021. *The Pivotal Generation: Why We Have a Moral Responsibility to Slow Climate Change Right Now*. Princeton and Oxford: Princeton University Press.
Stemplowska, Zofia. 2016. Doing More Than One’s Fair Share. *Critical Review of International Social and Political Philosophy* 9: 591–608.
Stockton, Ben, and Lawrence Carter. 2023. “UAE Oil Company Employees Given Roles in Office Hosting Cop28.” *The Guardian*, February 3.
Supran, Geoffrey, and Naomi Oreskes. 2021. Rhetoric and Frame Analysis of ExxonMobil’s Climate Change Communications. *One Earth* 4: 696–719.
Tabuchi, Hiroko. 2020. How One Firm Drove Influence Campaigns Nationwide for Big Oil. *The New York Times*, November 11.
Tehran Times. 2019. Rouhani Says Separating Budget from Oil Requires ‘Precise Surgery’. *Tehran Times,* October 14. www.tehrantimes.com/news/441162/Rouhani-says-separating-budget-from-oil-requires-precise-surgery.
UNEP. 2022. *Emissions Gap Report 2022: The Closing Window — Climate Crisis Calls for Rapid Transformation of Societies*. Nairobi: United Nations Environment Programme. www.unep.org/emissions-gap-report-2022.
Vähäsarja, Sari. 2020. Kasvituotteiden menekki yllätti perinteisen lihatalon: Poutun Muu-tuotteiden pohjalta syntyi uusi yhtiö, joka tähtää liikevaihdon tuplaamiseen vuosittain. *Yle.fi*. October 22.

Watson, Robert, James J. McCarthy, Pablo Osvaldo Canziani, Nebojsa Nakicenovic, and Liliana Hisas. 2019. *The Truth Behind the Climate Pledges*. The Universal Ecological Fund (Fundación Ecológica Universal FEU-US).
Wenar, Leif. 2018. Beyond Blood Oil. In *Beyond Blood Oil: Philosophy, Policy, and the Future*, ed. Laurie Shrage. Rowman & Littlefield.

CHAPTER 3

Responsibility as Members

Introduction

As I argued in Chap. 1, humans need to act with great urgency to mitigate climate change to a level to which societies can still adapt. This goal could be met through national regulatory means if an effective global emissions system was in place, or perhaps through market mechanisms if the social cost of carbon was internalised in prices. However, at the moment neither solution has materialised. While the previous chapter discussed the responsibility of collective agents when global collective solutions are for the most part still lacking or insufficient, this chapter discusses how members of collective agents should respond to the urgent need to mitigate climate change in light of these realities. I will argue that mediated responsibility as members of organised collectives is important and could often even be our best way of taking responsibility for climate change. My main message is that we should not forget this causally potent and individually salient responsibility, which rarely gets highlighted in climate ethics. In many cases, it might even be our most promising way of making an impact when it comes to large-scale collectively caused harms in which a collective solution is missing. After all, while not everyone is cut out to be an activist or a climate leader, nearly everyone is a member of a collective of some sort where they could play a part in effecting change. This is also the framing that we should push for in political philosophy and public debate alike to try to motivate people to act.

S. Hormio, *Taking Responsibility for Climate Change*,
https://doi.org/10.1007/978-3-031-51753-2_3

During the course of this chapter, I discuss what the responsibility of collective agents (discussed in the previous chapter) might entail for their members. While most of my examples will be about for-profit corporations, the paradigm example of collective agents, the general structure of the argument can be applied also to other collective agents and their members. These include universities, local governments, non-governmental organisations, art and sports associations, and so on. In terms of taking responsibility for climate change, it is important that members of all kinds of collective agents get on board.

I begin by discussing what members should do to try to get their collective to take climate mitigation seriously. After all, to a great extent it is the members that decide how collective agents operate (forces external to the collective, like legislation, influence the collective as well). After this, I will discuss how the members should assess their roles within these collectives to see where they could make a difference. I will also look at the impact of collective policies on the responsibility of their members. I end the chapter by looking at the importance of social activism, which is about working towards effecting change on a social issue even when that goal is not directly under your influence or control. The end of the chapter pays special attention to employee activism.

Members of Organised Collectives

As collective solutions to climate change are still missing for the most part, the role of individuals comes to the fore. We need an all-hands-on-deck approach in climate policies, as I argued in the previous chapter, which also means there should be action within various organised collectives to move their course. After all, collective agents are not some pre-destined machines that go towards their set goals in a steady matter. Instead, they are micro-societies in which members can often make a difference, especially in smaller settings. Collective agents of course vary widely in how democratic or autocratic they are, who gets to decide on what, and how rigid a hierarchy people enter when they take on a role. I discuss these issues later on in the chapter. There is also a lot of variation between collective agents not only in the constrains brought by collective procedures, but also in membership size. After all, most collective agents are not some huge conglomerates or institutes with massive staff numbers, but instead much smaller operations. Regardless, my argument is that when looking at what collective agents should do, we should focus also on the human parts that they are made up of. In other words, we should not forget about the responsibility of individuals embedded in collective contexts.

As discussed in Chap. 1, organised groups are structured entities, in which members play the normatively bound roles embedded in the organisational structure (Ritchie 2020). It is important to keep in mind that collective agents, such as corporations, are not a homogenous group that respond to external stimuli, such as environmental regulation, in the same way (Dupuis and Schweizer 2019). Instead, each collective agent has their own character and ethos (Tuomela 2007). Research suggests not only that distinctive and persistent organisational cultures and ideologies exist but also that these reflect the ideologies of a wider pool of members than those in the narrow realm of executives and other elite members of collectives (Gupta et al. 2017). This culture and ideology influence the decision-making of the collective, including how responsive the collective is to a range of causes (Hormio forthcoming). Although individual members do not control the rational point of view of a collective agent (i.e., the set of beliefs and desires that guide its action), they can influence it through various means (Hess 2018). In addition to engaging with decision procedures, this can also take the form of helping or hindering collective decisions through the ways in which the members carry out their tasks. Members at various levels of the organisation can propose and endorse strategic initiatives, while the top executives attend at least to some degree to the prevailing belief systems of their organisation, regardless of their personal beliefs.[1] Furthermore, organisations tend to attract, select, and retain like-minded employees, therefore gaining more support for their ideology and culture (Gupta et al. 2017). All in all, individual members are affected by *and* affect the collective culture and ethos.

Who, then, are the members of an organised collective? This naturally depends on the rules and organisational set-up of that particular collective, as the rules and conventions of membership vary, but in general members are those individuals who constitute the collective in question (Miller 2020). In universities, for example, there is academic and administrative staff, undergraduate and postgraduate students, janitorial staff, chancellor or president, heads of faculties, trustees, and so on. Universities, businesses, non-governmental organisations, associations, and other collective agents usually have rules and decision-making procedures in place about

[1] This is done to enhance their own legitimacy and to increase the acceptance of their decisions.

membership. Some roles come with more room to act and to make a difference than others. After all, responsibility *qua* members is not only affected by what you are a member of, but more importantly what your role within that collective is.

Members of collective agents have differentiated roles and responsibilities, and power. In most corporations, for example, the board of trustees wields a lot of power. Changes of even single individuals in the company board can have a big influence on the values of the corporation (Dupuis and Schweizer 2019). Executives are another example of key members. In organised collectives with clear role definitions, executive roles "are defined by their authority to make or contribute to key organizational decisions", whereas employees whose positions are delineated in terms of tasks have less influence as they mostly carry out the daily business of the organisation necessitated by its structures, policies, and the decisions taken by executives (Isaacs 2011, 104–105). The people who hold power within the collective can be described as the core agents. Core agents usually bear the greatest responsibility as they have the most power and are involved in drafting the policies, procedures, and goals for the collective. This means that they also have the most knowledge about the same. In contrast, peripheral agents have much less power in the organisation, or knowledge about collective goals and commitments. They are usually also less invested in the collective action and outcomes.

Many of us fall somewhere between core and peripheral agents. We have a say in shaping our roles, but we are not free to act as we wish, because our roles constrain us to a degree (although they should not prescribe what actions we should take, something I will discuss soon). Importantly, our roles also enable us to do things we could not do otherwise. By acting as members of collective agents, we often employ some of the power of the collective. Responsibility *qua* members therefore cannot be discussed separately from role responsibilities as they overlap. Roles can give us access to collective power not otherwise available to us as individuals. Once an individual occupies a role in an organisation, they also acquire access to special powers and resources (Zheng 2018). Thus, the core agents with more power and knowledge will always have stronger responsibility *qua* members of organised collectives than the peripheral agents. However, it is important to keep in mind that while those individuals might be peripheral within a particular collective, they might be able to influence some other collective agent through their membership in that collective. After all, many of us are members of various collective agents also outside our work roles.

Although our responsibility as members is enabled and shaped by the roles we occupy, this concept is still distinct from role responsibility. Role responsibility is attached to the role and is defined by institutions and the collective in question. A president has certain role duties, constrained and enabled by the institution, regardless of who occupies the role (how well they fulfil these is another matter). Such responsibility is role-specific, not person-specific. In contrast, I argue that responsibility as members of organised collectives is attached to the person: it is person-specific and tied to your membership. While responsibility *qua* members is both enabled and shaped by our role in the collective—without membership in the collective, there would be no such responsibility—it is not *defined* by the collective. Rather, responsibility as members is a matter of moral responsibility, how we should act towards each other in the world we live in. In other words, responsibility of members goes beyond what is involved in their designated, contractual roles (Collins 2019, 193–194). The kind of membership responsibilities I discuss in this chapter derive from and are part of the wider moral considerations that are mostly formed outside the confines of the collective norms and procedures. This means that sometimes we should even act against our roles, like when we are asked to do things morality would advice against, or asked to stay silent on issues that should be discussed (I discuss employee activism towards the end of this chapter). Still, the point remains that our roles as members of collective agents create special opportunities to take effective action.

As I argued in the previous chapter, collective agents which have the capacity to act on climate mitigation should act. If the leaders fail to enact policies and procedures for the collective to act in mitigating ways, this means that the other members have responsibility to try to enact change. If a collective agent is not doing enough (or not doing anything) about climate change mitigation, then it is up to its members to get the ball rolling, so to speak. Looking at your role from your viewpoint is the first step. I suggest that we have:

Climate change mitigation responsibility *qua* members of organised collectives

We should assess our role within organised collectives to see where we could make a difference and then push the collective to adopt policies and procedures that reduce its net carbon emissions or otherwise promote mitigation.

Not everyone has to act, but enough have to in order to make the necessary changes to ways of working. In some very small collectives, like an association consisting of three people, it could be enough if one member initiates the changes. However, within large collective agents, usually more are needed. It could mean just a handful of members, or it could mean that hundreds (even thousands) of people need to get together. It all depends on the situation, including the size and type of the collective agent (a large international non-governmental organisation, a middle-sized university, a small company, etc.). It also depends on the members who take the lead: their personal qualities, as well as their position and power within the collective.

We should assess our role within organised collectives to see where we could make a difference, and then try to push for the collective to take action on mitigation. The action that we push for does not have to be the most ambitious option. Sometimes small victories can help to set off bigger changes. Nor does the action have to be top-down mandated policy changes, but can take softer forms of impacting ways of working. Think of the social norms in academia that makes regular air travel not only normal, but also in many cases expected (networking opportunities afforded only through conferences, expectations of mobility by funders, etc.). It could be that a senior academic might have more impact on emissions by pushing to change these norms in their academic field, rather than engaging in other means, like lobbying the university administration to make changes in their procurement practices, for example. The context is important when making the call about which action to push for, and in how it should be pushed for. The wider context needs to be taken into account as well. In the case of academia, that means trying to balance the genuinely international nature of scientific research and the benefits of in-person exchanges of ideas with the urgent need to curb emissions from aviation.

Responsibility *Qua* Members of Organised Collectives

In this section, I discuss what it means to take seriously your climate mitigation responsibility as a member of an organised collective. Although my focus in this chapter is on corporations and smaller companies, this basic structure applies to other collective agents too. I suggest that a failure of a collective agent to act on an important social issue, despite having the capacity to do so, gives its members a reason to try to push the collective agent to take action. I also discuss threshold goods, the possibility of systemic failures, and elaborate on the idea of assessing our roles within a collective. Members, such as employees and shareholders, should look at their roles within organised collectives to see where they could make a difference from their special vantage point. The overall idea is not that we should do this all of the time and in relation to all of our roles, but enough in the wider context that includes also our other responsibilities and other ways that we might already be taking pro-mitigation action. As said in Chap. 1, the goal is climate sanity, not sainthood.

Given the importance of the range of roles discussed earlier, what is it that individual members should do in light of their responsibility as members? There will not be a one-size-fits-all answer to the question, as the responsibilities of members are not alike in different kinds of organised collectives. While in small, egalitarian collective agents there might not be much difference between members in terms of the power that they wield, this is not the case in large and hierarchical collectives. In the latter, the core members and those who are operative members in key areas are the ones who should primarily take action. If enough support the mitigation option, it will be implemented by the collective. In most cases, I suggest that we are talking about threshold goods (Lawford-Smith 2012), meaning that enough members must take part in the action to ensure a certain outcome is brought about. To make decisions such as switching energy suppliers of a retail chain, or reducing the number of flights an association takes annually, the operating members usually have to take a vote or engage in another decision procedure, depending on the collective. When the organised collective grows in size and complexity, there will be other factors that come into play, like additional bureaucracy, which might make implementing any kind of change more difficult, but that does not dilute the central idea of a threshold good (i.e., that a member must act unless enough other members already do). The responsibility of an individual member in such circumstances would be to show support for the mitigation option in one way or another. Hess (2018) argues that members at every level of a corporation are likely to have some mechanism at their disposal with which to either hasten or resist changes to corporate positions.

There could be an obvious mitigation path for the collective to take, such as adopting more carbon-neutral production methods, but this need not be the case. After all, an organisation can help to mitigate climate change in multiple ways, such as through reducing emissions, investing in green energy, making changes in their products or services, engaging in lobbying, and so on. Let us say that for some particular collective agent, all of these options are equally good. That is, they require an equal amount of effort and resources and are projected to produce similar enough results. Let us further stipulate that the collective agent is engaged in activities that produce a lot of GHG emissions, but has the means to act differently. In a case like this, there is no obvious option to take, but it is still clear that *some* course of mitigation action should be chosen. In a hierarchical organised collective, there will be key operative members who should set the ball rolling. In most cases, this will mean those who hold leadership positions, including the board of trustees. Many times, the shareholders hold the most power (sometimes, they can be too dispersed as a group for that). But if all these people fail to do what they should, some capacity-relative responsibility can fall on the other members to at least try to effect change.

Whenever there is a collective goal or a common end that can only be achieved if enough individuals work together, questions of fair share of burdens arise. As mentioned in the previous chapter, there is a lot of debate around whether agents are required to take up the slack left by the failure of others to do their fair shares in achieving a collective good. Some argue that agents are never required to do more than what is considered as their initial fair share, while others claim that considerations of fairness should not come into determining what should be done in the light of noncompliance by others. Recall that Zofia Stemplowska (2016) argues that we should prioritise the possibly dire needs of the objects of the duty over the reasonable extra costs incurred by those who take up the slack. In this chapter, I make a narrower claim within organised collectives. Namely that if the collective agent, taken as a whole, has the capacity to act and should act, but its leadership fails to do so, then some of the responsibility to act falls on the other members. In other words, they should try to push the collective agent to take action (within the parameters that it is not overdemanding). The failure of a collective agent to act on an important social issue, despite having the capacity to do so, gives its members a *pro tanto* reason to try to push the collective to take action. In assessing if they should do so *all things considered*, they should assess their position, networks, personal qualities and capacities, as well as who else is willing to join in. Social equality or inequality within the members also matter. For example, one could refuse to take action if this would deepen an already existing relational inequality with those with more power who do not want to do their share (Bajaj 2019).

It might also be that trying to push the collective agent to take action is impossible for the members: they just do not have the channels within the organisation to do so. Going outside the organisation to raise alarm about the issues could be too demanding, especially if there is a high likelihood of repercussions, like getting fired. I will return to this in the next section. That being said, it should also be noted that most collective agents are not some corporate conglomerates or large institutions like universities, or even nation-wide retail chains. Instead, many collective agents are small- and medium-size companies, associations etc. These consist of maybe only a handful of people who know each other or of smaller subgroups/teams who have a rough knowledge of the other subgroups and personal relations to at least some of them. In such collectives, there will still be hierarchy and rules, but even without a democratic set-up, members will have countless opportunities to influence each other and the collective ethos through formal and informal means.

Now, when the collective has not decided on a strategy to address climate change, it is very likely that the collective will not take effective mitigation action until such a strategy is in place.[2] Non-executive members could plausibly demand those in charge to set the strategy, and if they do not respond, make the issue known inside the collective (possibly outside as well to add to the pressure). This would, at this stage, be all that could be demanded of the ordinary members. This is because in large, hierarchical, non-egalitarian collectives, the content of the collective goal must be settled before individuals can go about taking care of their share. Alternatively, if the best course of action is somehow obvious and something that can be started incrementally, a group of members can initiate action, thereby making it the *de facto* choice. Depending on exactly how hierarchical the collective is, there might be some members who are in practice excluded from having a capacity-related share, for example, members who are relegated to the margins and who have little or no real power in terms of their roles. Saying that, the overall structure of membership

[2] For the sake of simplicity, I will assume here that the collective agent operates in an environment in which it is widely acknowledged that all actors should do what they feasibly can to mitigate climate change, but in which no legal framework or enforcement measures are in place.

responsibilities is that members should do what they can to push for action, as long as this is fitting and not overly demanding. I agree with Stephanie Collins (2019, 198) that if a collective agent has a duty to do something, but has no procedures in place to do so, it implies that its members should use their roles (if possible and appropriate) to put suitable inputs into the collective's decision-making procedures towards that goal.

If and when a collective agent has a mitigation plan, then we are less likely to encounter the problem of wondering what one can do. I am assuming that the collective has done the planning properly and delegated tasks to individuals, thus helping to define each member's part. If the responsibility to take meaningful mitigation action is delegated to a particular team, then the only capacity-relative share that remains for others outside that team is to keep an eye on things to the extent that they see that progress is being made. It will fall to leaders to ensure that *enough* progress is made. Responsibility as a member can be quite straightforward once a proper collective plan is in place. But what about when there is no plan, when have you discharged your responsibility? Say that you have pushed for a vote, but failed to get the majority onboard. If the vote was close, it seems clear that the members should try again, perhaps arming themselves with better arguments. But say that the vote clearly showed that not enough other members are willing to participate in bringing the threshold good about. At what point can you quit?

With mediated responsibility for collectively caused harms, I suggest that we should try to assess three things as members of organised collectives that contribute to the problem. These are 1) the scale and urgency of the harm or issue, 2) the causal role and capacity of the collective we are a member of in relation to that issue, and 3) our opportunities to make an impact within the collective. With climate change, the scale and urgency of the harm is given. The issue is therefore something we should all think about as members of collectives. Causal role suggests that members of collective agents that have high emissions, like those in heavy industry, should be more persistent in demanding action than members of those with lighter carbon footprints. This does not erase the need for all to push for their collectives to become carbon-neutral as soon as possible, nor does it mean that members of collective agents with light carbon footprints are off the hook. However, as we need drastic emission cuts soon to alter the current catastrophic course, the scale of the early wins matters. Last but not least, our role is important in determining how persistent we should be. This includes both our official role and our personal qualities.

If a systemic failure of some kind in the way that the collective operates prevents it from taking meaningful mitigation action, the members are responsible for fixing it.[3] With systemic failures, I have in mind cases in which there is a bad collective outcome despite it not being intended and without there being negligence (Hormio 2018). Rather, it is due to a design fault of some kind. Let us say that we have a collective agent that is engaged in building architecturally desirable houses for people with the means to invest in their homes. Let us further say that there are some easy switches that they could make in their plans to make the new houses 50% more energy-efficient. Even better, these switches do not cost them a lot and can be achieved without compromising on the design. However, due to some systemic fault in internal communication, the information never gets passed on to the relevant people in the collective. There is also no discussion at the level of the collective on making their activities as carbon-neutral as possible. The houses are built to be wasteful in terms of their energy usage. Now, some outsider points this out to several members of the collective. It then falls to the people in charge to remove the systemic failure (in communication, or elsewhere) that blocks the possibility of the collective being informed about the availability (and desirability) of such switches (and consequently forming a decision to take mitigation action).

But what if the collective is such that no one is in charge of fixing such systemic failures? Apart from this being a very poorly organised collective, this would still not block the responsibility to take action. In such a case, the collective should decide who will be tasked with removing the systemic failure. If they fail to do so, due to not caring enough about the failure, it becomes a case of negligence. If they fail to do so for other reasons, it would then become a case of direct responsibility for the executives and owners. After all, this kind of collective is no longer one that you could transfer any responsibility to, as it would not be able to take action in a responsive way. Or as Collins (2019) argues, members should take steps to overhaul things like the collective's decision-making procedure or its material components if these inevitably lead the collective to act in ways

[3] I include the discursive dilemmas discussed by Pettit (2009, 99–100) in this category: if a group agent has a decision structure in place that produces group decisions that each member individually rejects, this decision procedure has to be amended.

that are bad. All in all, when a collective agent does harm, it falls on its members to fix it or disband it.

Impact of Collective Policies on Members' Responsibility

I noted earlier that collectives can vary widely in how democratic or autocratic they are, and how rigid their internal hierarchy is. In this section, I want to discuss the impact of this variation in terms of what I have argued so far. I also discuss employee activism, a form of social activism, which can be defined "as instances in which individuals or groups of individuals who lack full access to institutionalized channels of influence engage in collective action to remedy a perceived social problem, or to promote or counter changes to the existing social order" (Briscoe and Gupta 2016, 674). Such social activism can also be undertaken by other members of a collective agent, not just its employees. One example of this is the university students leading the fossil fuel divestment movement in and outside university campuses (Bratman et al. 2016).

The more democratic a collective agent is, the freer its members are in expressing their opinions. In a democratic organised collective, all members have a say in collective decisions. The members get to voice their ideas and concerns more or less freely, despite of their rank or role. It might not be an equal say, but a say nonetheless. This could be achieved by debate and direct voting in smaller collectives. In a larger collective, what is probably needed is a combination of electing representatives among the members and utilising participatory methods when needed (see Chap. 6 for discussion on participatory democracy). We are used to thinking this way about certain associations and institutions, but it might sound strange when it comes to companies and corporations. Even so, some theorists argue that as organisations, corporations are comparable to the state, and therefore they too ought to be ruled democratically (Landemore and Ferreras 2016). I believe that corporations should become more democratic, for example, by electing worker-representatives to the company board.[4] Of course, internal democracy is not the only

[4] What I have in mind are measures like those found in the Accountable Capitalism Act, a bill introduced by US Senator Elizabeth Warren in 2018. Following in the footsteps of Germany, the bill suggests that companies should select at least 40% of directors from among their employees.

thing that matters, as how free people are to express their opinions in the wider society also counts, as well as how secure they feel in social safety nets or gaining alternative employment if necessary, for example.

Two scenarios need to be distinguished. On the one hand, we have collective agents that do not have any policies in place to facilitate discussion and organisational innovation around mitigation. On the other, we have collective agents that have policies to actively silence discussion around climate change. This could be for ideological or economic reasons, or both. I will illustrate these differences by discussing the responsibility of the employees of two invented large concrete manufacturers. Let us say that both Concrete Company and its competitor Certainly Cement, our two invented companies, are not interested in riding the green industry heat wave (see the previous chapter on the need to decarbonise heavy industry). Rather than embracing the future, their leaders have adopted a policy of wait-and-see with regard to reducing their emissions. They have the means—both companies could afford to invest in carbon sequestration and ultra-high concentrated solar—but they lack the will.

The two collective agents differ starkly in their policies regarding internal debates. Concrete Company does not have policies in place to facilitate discussion and organisational innovation around attaining carbon neutrality. It is not that the leaders are against mitigation, they simply do not view Concrete Company as an agent of change, preferring to adopt new technologies only when required by law or market realities. In contrast, Certainly Cement is positively against any talk of going carbon-neutral. The majority owners of the company believe that climate change is, if not a hoax invented by enemies of capitalism, then at least vastly exaggerated as a threat. As a consequence, climate change has become something of a taboo among the leaders and the employees. Not only is no internal debate around climate change facilitated, but the issue is also never mentioned even in passing in annual reports or other company communications. Discussions around climate change are nipped in the bud and repeat offenders are penalised by being overlooked for promotion. In other words, Certainly Cement has informal policies to silence discussion around climate change.

When it comes to Concrete Company, individual employees or shareholders should try to bring mitigation into the internal discussions, perhaps by sharing news about the latest technological solutions. The first step is to put carbon neutrality on the company's internal agenda, to nudge members to think about the issue from Concrete Company's viewpoint. Internal discussion and pressure about the need to transfer to new technologies are needed to get the issue incorporated into company policy. Employee activism can be conceptualised as 'insider activism', in contrast to the outsider activism practised by NGOs or social movements (Briscoe and Gupta 2016). With insider activism, there is usually more intimate access to internal company procedures and communications. Insider activists are thus often privy to more information than outsiders. Their activism can therefore be more targeted and potentially more effective.

With Certainly Cement, the employees face a much riskier road if they try to push for decarbonising action. With suppression of discussion, the members are effectively silenced. In these cases, the discussion could be taken outside the collective, for example, by tipping off a journalist or publishing an open letter with many signatories. Still, this is a big ask. Therefore, while employee activism in Concrete Company is covered by my formulation of what members should do, the situation at Certainly Cement could be too demanding for this to apply. That said, climate change is so well-known a risk these days that it would be strange if the shareholders of the company did not push for it to have a climate change business strategy in place. In cases of internal suppression of debate, shareholders could be the right members to question what is going on.

While pushing for action in Certainly Cement would be risky and falls outside my formulation of what members should do, there is a long tradition of people going above and beyond what can be demanded of them. The bravery of a few people can sometimes open up the floodgates on an issue that was previously brushed under the carpet, like what happened with #MeToo. The cost of activism comes down as more people take part. Raising concerns publicly can be about starting an internal conversation around a topic that has not been picked up to a sufficient degree by the management. While whistleblowing is traditionally marked by an insider revealing information that the organisation has entrusted them with (Davis 2005), modern collective whistleblowing is more akin to a group of insiders airing the dirty laundry of the organisation in public in order to protest its practices, often after internal channels have failed to take notice of their concerns. Some examples include employees objecting to Google technology being used for military purposes in Project Maven, or Microsoft workers protesting against the role of its technology in deporting and detaining immigrants in the United States under the company's contract with ICE. Similar activism has also taken place in relation to climate change. For example, Amazon Employees for Climate Justice want to see Amazon stop funding politicians and lobbyists who deny climate change and the aim to achieve zero-emissions logistics by 2030, along with ending contracts and custom solutions for fossil fuel companies. They argue that:

> As employees at one of the largest and most powerful companies in the world, our role in facing the climate crisis is to ensure our company is leading on climate, not following. We have to take responsibility for the impact that our business has on the planet and on people. (Amazon Employees for Climate Justice 2019)

Another example is a letter signed by over 1100 employees of the large consultancy firm McKinsey. The letter states that McKinsey's inaction on—or perhaps even assistance with—its clients' emissions goes against the company's stated values, is a serious risk to the firm's reputation, and poses problems for its ability to attract and retain excellent employees. The consultancy firm has worked with BP, ExxonMobil, Gazprom, and Saudi Aramco among other large oil and gas companies, helping them to boost productivity, cut costs and increase profits (Forsythe and Bogdanich 2021). In their letter, the employees wrote that McKinsey has "a moral obligation to take action to influence our clients' emissions and to show the leadership our stakeholders expect of us". Such employee activism often takes place in smaller scale, though, meaning that personal costs can also be higher. In order to protest its very close ties with the fossil fuel industry, six employees of the PR agency Edelman resigned in 2015, including the executive vice president of corporate responsibility practice, who later described the firm as the most central agency in climate obstruction (Joselow 2021).

Employee activism can also cut across collective agents and focus on an industry instead. One recent example is Clean Creatives, a group that targets creatives working for PR consultants and advertising agencies (alongside the agencies themselves) to try to stop them helping fossil fuel industry actors to distort debate around climate action (Edgecliffe-Johnson 2022). The campaign asks young leaders and professionals in the ad and PR industry to co-sign open letters addressed to agency executives to cut ties with fossil fuel companies. It is run by a group of strategists, creatives, and industry leaders working in the field of ad and PR, so it is thus a form of peer-to-peer appeal. The campaign states on its website (cleancreatives.org/creatives): "We have the power to shape public opinion, and steer profits to the clients we work for. Let's do it on behalf of clients that are doing good, not fossil fuel companies." This is an example of a campaign that targets collective agents through direct appeals to their members.

To explain the surge in employee activism, commentators point towards the different values of the younger generations now in employment, some of whom are already in leadership positions (Coulman 2019). A benchmarking report by a whistleblower services provider shows that whistleblowing reports increased by over 50% from 2016 to 2018, suggesting that employees "feel more empowered to speak up—either by workplace or external influences" (Expolink 2019, 22). Once the traditional routes of flagging up grievances with company policies are exhausted (through HR and line managers), social media savvy Millennials make their grievances public through letters, petitions, or other action, like participating in Climate Strikes.[5] Another reason for the current bout of activism might be the lack of unions and protection for workers, with companies like Amazon viewing organised labour as incompatible with its "direct relationship with associates" (Bruder 2019, 68). This new wave of employee activism is marked by social change as the goal (rather than improvements to one's personal conditions) and collective whistleblowing through social media (Coulman 2019). The distinction between old and new need not be exaggerated, however. Whistleblowing in general happens when an organisation does not have adequate internal mechanisms in place to deal with the concerns of its members, or when these concerns have been met with silencing or indifference on the part of the management. Organisations can deal with whistleblower complaints effectively internally if there are good procedures in place for stakeholders to voice their concerns, anonymously or in confidence. It has been argued that organisations have a duty to set up mechanisms to address wrongdoing (Ceva and Bocchiola 2019),

[5] One survey in the USA found that 48% of Millennials see themselves as employee activists, which is significantly more than older generations: only 27% of Baby Boomers viewed themselves as activists in the workplace (Weber Shandwick 2019). Although categorising generations in this way is not an exact science, Pew Research Center suggests that Millennials are those born between 1981 and 1996, while people born from 1997 onwards comprise Gen Z (Dimock 2019). Worldwide, Millennials and Gen Z account for over half the population and most of the global workforce (Deloitte 2019, 27). In the US, those Millennials who identify as Republicans take climate change more seriously than older generation Republicans (Funk and Tyson 2020).

so the need for whistleblowing and collective activism can be seen as a failure of the collective agent and its internal procedures.

I hesitate to extend the call for workplace activism to situations in which the personal costs are potentially very high, such as losing your job, friends, career prospects, or incurring other financial and social penalties. Activism is usually not something that can be demanded of us: rather, it is something commendable, but up to the individual's discretion. I am not convinced we could come up with some neat formula that would tell us when these costs are too high. Much has been written about whistleblowing ethics (Ceva and Bocchiola 2020), which might offer some guidance to an individual facing a dilemma.[6] Still, it is up to the individual members to make such an assessment. Activism can have high individual costs even for those who are in relatively privileged jobs (at least in countries where employee protections are weak). For example, internal documents show that several Google employees have faced retaliation for protesting against the company's way of dealing with sexual harassment (Ghaffary 2019). It has been argued that members should develop routines for keeping the larger collective they belong to in check, like coming up with monitoring practices, and that they have member responsibility even when they bear no responsibility as the enactors of the collective action (Pettit 2009). This is true, but the costs have to be kept in mind. Having said that, there are certain cases when a person has to stand up for what they believe in, or to support their peers in their valid demands. How to formulate such demands and identify such situations in a simple formula eludes me (unless the formula runs along the lines of 'do x, unless the costs are too high', but this is hardly illuminating). If the situation clearly calls for action, perhaps sometimes the employee can only decide to quit, as long as that option is economically viable for them. This is, of course, far from ideal and far from the most productive option.

[6]This is not to say that in the real-world whistleblowing is necessarily preceded by an assessment of acceptable costs to oneself. Although whistleblowers are often praised for their integrity and bravery—and deservedly so—empirical research suggests that many whistleblowers do not know what they are getting into. Whistleblowing is often motivated by 'passionate attachment' to professional norms and the ideals of the organisation (Kenny et al. 2018). Rather than taking a calculated risk, the study found that many whistleblowers did not foresee the costs to themselves in reporting the transgressions they witnessed. This suggests that the risks associated with whistleblowing action are not always appreciated and that misplaced trust in the integrity of the collectives we serve can make for accidental whistleblowers.

Be that as it may, instances of recent employee and shareholder activism may be one reason to be hopeful. If enough people are not content to just go with the collective flow, and actively engage instead in a dialogue with the leaders of the collectives they are a part of, then maybe large-scale action on climate change is still possible in the short period of time that is (hopefully) still left before the climate crisis spirals out of control. But my main message has not concentrated on activism as such, but instead on willing to steer and push the collectives that we are members of towards solutions that help with mitigation. Often this can be done internally without the need to blow the whistle, through measures like innovation and collective brainstorming.

Concluding Remarks

I have argued that collective agents across the board should play their part in decarbonising our societies. Our mediated responsibility as members of organised collectives could be the key responsibility for many of us when it comes to collectively caused harms like climate change. This intermediate level of responsibility can be causally potent and individually salient for initiating action. As role-occupiers, individuals wield power that they do not have as individuals. As members of the same collective agent, they can come together to initiate change. Membership presents a potential opportunity for action.

While I have only discussed climate change action, the general structure can also be applied to other issues; that is, the argument about how members could—and should—try to get the collective actors they are a part of to act on important issues that the collective is currently passive about. In order for a collective agent to react and respond quickly to its environment, it needs members who are alert. Emerging challenges and changing social circumstances require existing policies to be amended or new solutions to be invented. In such cases, individuals should look at the influence and impact they can have when acting within their various roles.

References

Amazon Employees for Climate Justice. 2019. Amazon Employees Are Joining the Global Climate Walkout, 9/20. *Medium*, September 9.

Bajaj, Sameer. 2019. The Weight of Fairness. *Politics, Philosophy & Economics* 18: 386–402.

Bratman, Eve, Kate Brunette, Deidre C. Shelly, and Simon Nicholson. 2016. Justice Is the Goal: Divestment as Climate Change Resistance. *Journal of Environmental Studies and Sciences* 6: 677–690.

Briscoe, Forrest, and Abhinav Gupta. 2016. Social Activism in and Around Organizations. *The Academy of Management Annals* 10 (1): 671–727.

Bruder, Jessica. 2019. The Immigrants Who took on Amazon: How a Group of Somalis Became Leaders in the Fight to Change a Tech Behemoth. *Wired* December: 58–69.

Ceva, Emanuela, and Michele Bocchiola. 2019. *Is Whistleblowing a Duty?* Cambridge: Polity Press.

———. 2020. Theories of Whistleblowing. *Philosophy Compass* 15: e12642.

Collins, Stephanie. 2019. *Group Duties: Their Existence and Their Implications for Individuals*. Oxford: Oxford University Press.

Coulman, Lauren. 2019. Why Is Employee Activism on the Rise? *Forbes.com*, May 30.

Davis, Michael. 2005. Whistleblowing. In *The Oxford Handbook of Practical Ethics*, ed. LaFollette Hugh. Oxford University Press.

Deloitte. 2019. *The Deloitte Global Millennial Survey 2019: Societal Discord and Technological Transformation Create a "Generation Disrupted"*. Deloitte Touche Tohmatsu Limited.

Dimock, Michael. 2019). Defining Generations: Where Millennials End and Generation Z Begins. *Pew Research Center Fact Tank*, January 17. www.pewresearch.org/fact-tank/2019/01/17/where-millennials-end-and-generation-z-begins.

Dupuis, Johann, and Remi Schweizer. 2019. Climate Pushers or Symbolic Leaders? The Limits to Corporate Climate Leadership by Food Retailers. *Environmental Politics* 28 (1): 64–86.

Edgecliffe-Johnson, Andrew. 2022. Activists Target Public Relations Groups for Greenwashing Fossil Fuels. *Financial Times*, January 11.

Expolink. 2019. *Whistleblowing Benchmarking Report 2019*. Expolink Europe Ltd.

Forsythe, Michael, and Walt Bogdanich. 2021. At McKinsey, Widespread Furor over Work with Planet's Biggest Polluters. *The New York Times*, October 27.

Funk, Cary, and Alec Tyson. 2020. Millennial and Gen Z Republicans Stand Out from Their Elders on Climate and Energy Issues. *Pew Research Center*, June 24. www.pewresearch.org/?p=302449.

Ghaffary, Shirin. 2019. Dozens of Google Employees Say They Were Retaliated Against for Reporting Harassment. *Vox*, September 9.

Gupta, Abhinav, Forrest Briscoe, and Donald C. Hambrick. 2017. Red, Blue, and Purple Firms: Organizational Political Ideology and Corporate Social Responsibility. *Strategic Management Journal* 38: 1018–1040.

Hess, Kendy M. 2018. Who's Responsible? (It's Complicated.) Assigning Blame in the Wake of the Financial Crisis. *Midwest Studies In Philosophy* 42 (1): 133–155.

Hormio, Säde. 2018. Culpable Ignorance in Collective Setting. In *Action, Value and Metaphysics - Proceedings of the Philosophical Society of Finland Colloquium 2018 - Acta Philosophica Fennica 94*, ed. Jaakko Kuorikoski and Teemu Toppinen, 7–34.

———. forthcoming. Collective Agents as Moral Actors. In *Collective Responsibility: Perspectives on Political Philosophy from Social Ontology*, Studies in the Philosophy of Sociality, ed. Säde Hormio and Bill Wringe. Springer.

Isaacs, Tracy. 2011. *Moral Responsibility in Collective Contexts*. New York: Oxford University Press.

Joselow, Maxine. 2021. How Big Oil Relies on Some PR Firms to Block Climate Action. *The Climate 202, The Washington Post*, November 30.

Kenny, Kate, Marianna Fotaki, and Wim Vandekerckhove. 2018. Whistleblower Subjectivities: Organization and Passionate Attachment. *Organization Studies* 41 (3): 323–343.

Landemore, Hélène, and Isabelle Ferreras. 2016. In Defense of Workplace Democracy: Towards a Justification of the Firm–State Analogy. *Political Theory* 44 (1): 53–81.

Lawford-Smith, Holly. 2012. The Feasibility of Collectives' Actions. *Australasian Journal of Philosophy* 90 (3): 453–467.

Miller, Seumas. 2020. Collective Moral Responsibility as Joint Moral Responsibility. In *The Routledge Handbook of Collective Responsibility*, ed. Saba Bazargan-Forward and Deborah Tollefsen, 38–50. New York: Routledge.

Pettit, Philip. 2009. Responsibility Incorporated. *Rechtsfilosofie en Rechtstheorie* 38: 90–117.

Ritchie, Katherine. 2020. Minimal Cooperation and Group Role. In *Minimal Cooperation and Shared Agency*, ed. Anika Fiebich, 93–109. Studies in the Philosophy of Sociality: Springer.

Stemplowska, Zofia. 2016. Doing More Than One's Fair Share. *Critical Review of International Social and Political Philosophy* 9: 591–608.

Tuomela, Raimo. 2007. *The Philosophy of Sociality: The Shared Point of View*. New York: Oxford University Press.

Weber Shandwick. 2019. *Employee Activism in the Age of Purpose: Employees (UP) Rising*. Weber Shandwick, May 29. www.webershandwick.com/news/employee-activism-age-of-purpose.

Zheng, Robin. 2018. What is My Role in Changing the System? A New Model of Responsibility for Structural Injustice. *Ethical Theory and Moral Practice* 21: 869–885.

CHAPTER 4

Shared Social Orientation and Responsibility as Constituents

Unorganised Collectives

Collectives of various stripes bear responsibility for climate change, be they states, international bodies, corporations, or other collective entities. In the previous chapters, I focussed on organised collectives, arguing that collective agents of all kinds should take climate change into consideration in their operations. Nevertheless, climate change can be construed as a question of collective responsibility not only from the viewpoint of collective agents bearing responsibility for it, but also from the viewpoint of how the problem itself is inherently collective. Therefore, in discussions on collective responsibility for climate change, 'collective' can refer to the entity causing the harm (and/or bearing responsibility for it), or to the problem of climate change itself (Hormio 2023). This second way to conceptualise climate change as a question of collective responsibility focuses on how it is a harm that has been caused collectively and can only be solved collectively. *Collective action problem* is an aggregation of individual actions that produce an outcome not intended at the level of an individual action. It cannot be solved by any one agent acting unilaterally because remedial action must be enacted and supported by numerous agents.

Collective action problems like climate change raise questions of shared responsibility. We contribute to collective outcomes not only as members of collective agents (see Chap. 3), but also as constituents of unorganised collectives, that is, aggregates of people without an established

S. Hormio, *Taking Responsibility for Climate Change*,
https://doi.org/10.1007/978-3-031-51753-2_4

decision-making structure. I use the word 'constituent' (Hormio 2017), because there are no differentiated roles, or rules for entering and exiting membership in the way that there are for organised collectives. Instead, the unorganised collectives that are interesting in terms of moral responsibility are a set of individuals who are picked out by some normatively relevant fact. With the collective harm of climate change, the normatively relevant fact could be about things like the ecological footprint linked to one's consuming habits (e.g., 'daily consumers of dairy products' or 'frequent fliers'). The normatively relevant fact could also be about the potential of the constituents to do something good together, like 'local government procurement officials across the country'. Other unorganised collectives relevant for climate change responsibility include aggregates such as 'livestock farmers' and 'PR professionals helping the fossil fuel industry'.

This chapter focuses on shared responsibility as constituents of unorganised collectives. What is common for morally interesting unorganised collectives is that their activity is interdependent and based on shared values to some degree. Constituents can sometimes simply opt out, but this is easier with some unorganised collectives than others. One can quite easily stop consuming dairy every day, but changes that require rethinking one's professional life or livelihood is a much taller order. In these cases, what is called for is collective awareness raising about harmful practices and possibly changing these the norms around these with other constituents. Such awareness raising should go beyond simply sharing information about the impacts of harmful practices. Instead, it should also encompass the questioning of existing practices and norms, as well as suggestions of collective moral narratives. Sometimes creation of new collective agents is the best or the only way to tackle the issues relevant to the unorganised group.

Unorganised collectives differ on how much existing social relations there are between the constituents. By such relations I refer to either direct but loose personal ties between constituents (e.g., through some social media setting) or, more often, a degree of interconnectedness, for example through membership in collective agents operating in the same industry (such as people working in the same line of work, like 'hairdressers' or 'coal miners'). The stronger or more plentiful the existing ties, the easier it is—in theory at least—to start (organising) the required action to address the shared responsibility as constituents. On the other side of the coin, opting out might be socially costly. An example could be a former cattle rancher, now producing organic vegetables, whose neighbours stopped talking to them (for such a real-life story, see Bearne 2022).

In contrast, on the less interconnected end of the spectrum we have unorganised collectives that are just aggregates of people picked out by a normatively relevant feature, like something about their consumption habits. In such unorganised collectives composed of individuals with no existing ties or means of communicating with each other, it is harder to establish discussion on shared values. This is not impossible by any means, especially in the era of the social media, but the shared responsibility is more latent for such constituents. Saying that, when the constituents are non-interconnected in terms of personal relations, it is usually socially less costly for an individual to opt out, compared to unorganised collectives with existing social relations, however loose they might be.

In this chapter I look at shared responsibility for the impacts that our actions can have in aggregate with other such actions as constituents of unorganised collectives. Instead of trying to calculate the effects of small choices that we make in our everyday lives, I will suggest that a more fruitful approach is to focus on what I will label the *shared social orientation*. This is what constituents of all morally interesting unorganised collectives have in common. Shared social orientation offers a systemic viewpoint on shared responsibility, which emphasises collective level outcomes and the integrity of individual agents instead of the direct effects of their choices. One example is the aggregate impact we have on emissions as consumers. As noted in Chap. 1, part of our consumption falls under collective level structures, meaning that in some cases there might be no direct small choice or no direct small effect for the individual consumer. I argue that the reason why we should care about small contributions to collective outcomes should be spelled out in terms of shared responsibility. I discuss what this means in practice and how it can motivate us to care about collective harms.

The general idea of this chapter is that while most of the pressure to do better should be on collective agents, unorganised collectives should not be left out of the responsibility picture either. They can help us to appreciate the various structures and systems that we all are part of and how we are entangled and involved in upholding and recreating these. Furthermore, our shared responsibility *qua* constituents has the potential to be decisive in whether some action is taken or not by a collective agent, through a set of actions that can signal public acceptance or support for a policy or as a form of political support from the grass roots. When awareness is raised of the potential of the constituents to do something good together, this can result in positive things like new social movements or forms of activism.

Shared Social Orientation (SSO)

In this section, I explain what shared social orientation (SSO) is and how it can help us to conceptualise our responsibility as constituents of unorganised collectives. I suggest that the framing is a more fruitful alternative to an individualistic starting position in looking at collective action problems and our collective lives more generally. Although partly inspired by the focus on structures in Young's (2011) social connection model (see Chap. 1), my account is mostly based on Christopher Kutz's (2000) conceptualisation of individual contributions to outcomes by unorganised collectives (which he calls unstructured collectives).[1] My account expands the original model and applies it to climate change. Whereas Kutz uses the term 'quasi-participatory intention', I introduce the new term—shared social orientation—as I find it is more descriptive of the central idea behind individual contributions in unorganised collectives, in which we are often constrained by structures, but also help to shape each other's actions and values.

Our emitting actions take place in enduring collective contexts. The importance of structures was discussed in Chap. 1. With climate change, what is happening now and will happen in the future is linked to what has taken place in the past. Not all cases of responsibility for unorganised collectives have this feature. Sometimes circumstances can make it so that we must pull together to prevent some harm from being realised. An example is when two children have ventured too deep into water and a group of strangers on a beach need to work together to prevent them from drowning. In such cases, individuals who do not know each other should cooperate despite not having an established decision-making mechanism. The literature on circumstantial responsibility of such a random group has centred on the ability of the individuals to form a group agent capable of undertaking the action that the situation requires. Virginia Held (1970, 476) has argued influentially that a random collective (like passengers in a train carriage) can be held responsible for failing to act in cases when the action called for "is obvious to the reasonable person and when the

[1] For a suggestion on how to reconcile Young's criticism with Kutz's account, see Hormio (2017).

expected outcome of the action is clearly favourable." A failure to cooperate to prevent the harm counts as a blameworthy omission.

In cases of responsibility of such random, unorganised collectives, the individuals find themselves in the situation for reasons no more than bad luck. The composition of the collection of individuals is down to matters of geography and timing. This is not the case with human-induced climate change: the situation is fundamentally different from classic beach rescue scenarios. Climate change is not akin to cases in which collectives consisting of strangers should act together to prevent some harm that will predictably happen otherwise. We are not bystanders who should help the drowning just because we can; the circumstances of climate harms do not just present themselves to us. Although we inherited the problem from previous generations, we did not just happen to come across it. Rather, the situation is one in which patterns of existing joint action, and collective structures and institutions, set the scene for the actions that in aggregate cause the harm. Indeed, certain aspects of climate change can be conceptualised as structural injustices, as I argued in Chap. 1. When the threat or the harm is tied to structures, it is not enough that at the moment of danger we cooperate to prevent the harm. Instead, we need to fix the structures that create and facilitate the harm in the first place. To do so, we need to become aware of our positions within the structures. If we approach the polluters responsible for fossil fuel emissions as just an aggregate of individuals, we are failing to look into the structural issues that underpin climate harms. Such harms are more complicated for morality as the harmful structures are upheld or created by us.

When it comes to SSO, the actions of constituents are interdependent. They are enabled and constrained by structures, or based on shared values to some degree. I suggest that what we should try to do as constituents of such unorganised collectives is to pay attention to the harmful structures that we might be part of. More specifically, we should try to become aware of our social positions within these structures and be willing to take action to make changes for the better, should such an opportunity arise. To elucidate what I have in mind, I will first explain what I mean by a social position. We all occupy innumerable, partly overlapping social positions, which emerge from our material realities, values, desires, choices, and also from those of others. We are not aware of most of our social positions, because they can be so ingrained or mundane, that we don't pay attention to them. Most of them we also never become aware of because of the social conditions around us often make us blind to structures. As with anything social, our social positions are not set in stone. They can change either because we actively push ourselves towards new social position, or because things around us change. Sometimes this can provide previously non-important social positions new normative significance.

People who are in the same or similar enough social position in relation to some normatively important fact have a shared social orientation. Social position is a situation from which to examine the world and to (try to) affect it. Not all aggregates of people with a shared characteristic qualify as social positions, for example 'people with freckles'. Realising that I have freckles on my nose gives me no position from which to affect anything, as having freckles has no impact on the world, in aggregate or otherwise. As long as a property is socially neutral, having it does not give one any special social insight or viewpoint one would not otherwise have, because you are not treated by others in some specific way due to that feature, nor do you act in some specific way because of it. 'People with freckles' is thus not a social position and people who have freckles do not have a SSO, unless having freckles becomes somehow normatively significant in a society (usually caused by normative relevance placed from outside, like is the case with racism). SSO can also emerge from social kinds. Social kind is a collection of phenomena defined by a social property or feature (Ásta 2018), for example 'money', 'refugee' or 'porter'. Setting aside inanimate phenomena like money, some social kinds can also be social positions, like refugees or porters or vegans. Still, social position is a concept that is broader than a social role (e.g., 'a sibling', 'a student', 'a grandparent'), as it is also covers positions we do not necessarily have established expectations about, such as 'vegan influencer', or positions that would be too thin to be described as roles, such as 'daily consumer of dairy'. However, social roles can be a part of the building blocks for SSO, for example, 'climate parent' to describe parents who campaign on climate action.

SSO is about shared responsibility *qua* constituents of an unorganised collective. As SSO can be applied also to other systemic issues, not just to climate change, I will illustrate what I mean with an example from the following imaginary app.

Catwalk Closet

Customers can rent clothes through the app Catwalk Closet for a few weeks at a time, giving them access to a wider and fancier wardrobe than their budgets would otherwise allow. Many are attracted to the app for environmental reasons, as party gowns and other items that might be worn only once or twice are reused. Catwalk Closet advertises its credentials as an ethical company that does not discriminate when hiring new staff and pays the storefront clerks above minimum wages. What happens behind the scenes in the dispatching warehouses is a different story. Not only are warehouse workers paid a pittance at Catwalk Closet, but the working conditions are also cramped, and health and safety mostly ignored. Catwalk Closet uses harsh chemicals when cleaning the clothes, putting its workers in danger from the fumes. Several warehouse workers have developed debilitating lung conditions.

Catwalk Closet customers are an unorganised collective with SSO: set of people picked out because they have taken out a monthly subscription to the app. Their activity of borrowing and returning clothes is interdependent of others participating in such actions (it would not make business sense to have the app for only a few individuals). The customers are also likely to share some values, like a keen interest in fashion or a wish to reduce their environmental footprint. With the prevalence of social media, many of them could also openly identify as Catwalk Closet customers by tagging the company into their posts displaying their latest outfit.

Often our small contributions are made against an even looser collective structure than this, without a collective agent (Catwalk Closet Inc.) that the individual contributions revolve around and which clearly delineate the constituents. Most large-scale environmental harms are like this. Take microplastics in the ocean. There is no specific collective goal ("use loads of plastic!") to which individual agents contribute. Instead, what we have is "a set of individuals who jointly cause harm, against a background of interdependent activity and shared values" (Kutz 2000, 186). SSO in such cases can be things like valuing the inexpensiveness and convenience of plastic products, or even just being a consumer in a world filled with plastic. The latter is an example of a very loose and widely shared social position. We can unintentionally bring about harms through our interconnected actions that contribute to upholding structures that lead to harms like microplastics in the oceans or growth in CO_2 emissions. We should be aware—as much as is feasible—of the collective context in which we make our small contributions and our own social positions within it.

Let us say that a customer of Catwalk Closet reads an exposé of the situation in a local newspaper and is appalled. Recognising themselves as a part of the harm-contributing set can motivate customers to start demanding changes to the way the company operates. It might cause them to reflect on who they are, in the sense of a very loose group identification. After all, our identities are constantly influenced by our social environments to a degree, just as we also influence others in our sphere (May 1992, 173). More broadly, becoming aware of a shared social orientation can foster a sense of accountability for what we do together. I believe it can also open our eyes to the opportunities of what can be achieved together.

While he does not discuss climate change, Kutz does discuss ozone layer depletion and environmental harms more generally, describing environmental damage as "typically the result of the knowing but uncoordinated activity of disparate individuals, each of whose actions contributes only imperceptibly to the resulting harm" (Kutz 2000, 171). Note how the characterisation of environmental harms makes no reference to knowing or not knowing about the consequences: it simply states that the damage results from uncoordinated human activity. Few of us intend to pollute the environment: it is often the unintended effect of our other activities which sometimes we understand the mechanisms of, sometimes not. What is important is the collective level outcome that results from confluence of individual behaviour (Kutz 2000, 166).

There are two elements that, taken together, can account for why we can and should conceptualise individual acts as contributions towards climate change and other such collectively caused harms. The first is the systemic viewpoint, which identifies the set of agents who jointly cause harm (Kutz 2000, 186). Labels such as 'car owners' or 'daily consumers of meat' abstract away from individual agents towards the unorganised collective and this can help to orient ourselves socially in the big picture. They also draw focus on the systemic economic and social contexts that these behaviours take place in, such as transport infrastructure or the workings of the meat industry. Not only does the systemic viewpoint pick out a set of people, but it also paints them against the background of the collective systems feeding and enabling these behaviours. After all, as Kutz (2000, 167) suggests, the harms caused by unorganised collectives arise usually "in contexts in which deeper, systemic, forms of collective action lie".

The second element is character-based accountability based on symbolic reasons (Kutz 2000, 190–191). Symbolic reasons are not moral demands as such and supererogatory actions do not fit neatly with morality focused just on ethics, but they fall within the broader ethics of value.

Our actions might have no actual impact, but they symbolise what we value, tolerate, and are willing to be linked to. This has a lot in common with Bernard Williams's (1981) argument about integrity: we should not be fragmented agents, but we should be internally coherent. Our individual actions should be coherent with the values and collective actions that we promote. We might sometimes fail, and we will, but we should aim at things that correspond with what we value.

Any account of character-based accountability even partly based on symbolic reasons raises questions about integrity. Although integrity is a virtue that is cited frequently in ethics, it lacks precision as a term (Audi and Murphy 2006; Hourdequin 2010). In explaining why we should care about our small contributions even if they do not necessarily have an impact, I have evoked the narratives that the patterns that they form can tell us about our characters. This is not about some high-minded integrity and trying to keep one's hands clean at any cost. Instead, the reluctance of agents to participate in bringing about a harm is focused on the collective outcome and how it goes against one's values. It has been suggested in empirical work that a certain amount of emotional discomfort associated with ethical dissonance between your actions and your values can act as motivation to take better action, after a period of reflection (Brouwer et al. 2022).

My suggestion is that sometimes what is needed might be that we start feel bit uncomfortable with how things are. Marion Hourdequin (2010) argues that to understand an individual's obligations regarding climate change, we need both 'integrality' (internalisation of certain commitments) and 'integration' (unity among commitments). Human psychology does not lend itself to stark separation between personal and political obligations. She suggests that a person with real concern about climate change should make some effort towards realising social change, but also avoid frivolous emissions (such as Sunday joyrides with an SUV), as they should try to harmonise their commitments in all the spheres that they inhabit as much as they can. SUVs are sports utility vehicles that usually incorporate off-road features, such as four-wheel drive and high ground clearance. On my account, the social positions and patterns of behaviour revealed by SSO can tell narratives about our characters. The reluctance of agents to participate in bringing about a harm is focused on the collective outcome and how it goes against one's values. The reluctance to drive an SUV would thus be based on not wanting to be part of the set of drivers endorsing a shift to heavier vehicles that increase CO_2 emissions.

When a systemic issue becomes individualised, the structural aspects can get blurred or even fade completely into the background, meaning that the overall picture is also lost in terms of responsibility. I believe that awareness of our embeddedness in collective contexts helps us to make sense of our moral lives and responsibility in our interdependent world, where global issues like climate change abound. Shared social orientation offers a prism through which to frame, discuss, and debate questions of collective responsibility. Individuals have the power to wake up others to their SSO. Once this happens, changes at the collective, systemic level become possible. Changes in social norms can also make it easier for us to act in climate friendly ways and to reduce the demandingness on ordinary people over such action (Lichtenberg 2014). I believe that when we become aware of our SSO, change can happen even from unlikely sources. Pre-existing social identifications, roles and structures can motivate us to act when we find out that we are contributing to a harm (Kutz 2000, 188–189). Social change is of course notoriously difficult to predict. It is not necessarily a gradual process, and it does not need to be a conscious one either, as change can occur suddenly or even spontaneously (Andrighetto and Vriens 2023). Often social change is led by a minority motivated by moral concerns. Moral innovators are among the first to deviate from the status quo and they can play a pivotal role in the transition to a more sustainable society, as their actions may encourage others to make similar choices (Bolderdijk et al. 2018). My suggestion is that becoming aware of SSO might also lead people to become moral innovators.

What SSO Means for Responsibility

The SSO account points towards the need to be alert to opportunities to make a difference within an unorganised collective. I think that an important feature of any integrity account has to be accepting that we are all in different situations and our responsibilities are contextual, so we are often best placed to assess what kind of action we can and should take. Depending on the context, sometimes this means being willing to join action that others have started *or* the willingness to join in if others start such action *or* the willingness to continue doing an activity that contributes to an overall benefit. At other times, it means trying to initiate change yourself. As social creatures, we help to shape the values and actions of others and we are in turn being shaped by their values and actions. We usually do not know how others will react until there is an open debate about an issue, so after becoming aware of our social position as part of the problematic structures, the next step is often to try to start such a debate. After all, even when many people are needed to affect change, it is individual people who need to get the ball rolling (similar to what I discussed in the previous chapter in relation to individual members within organised collectives).

Sometimes this can be done at a low personal cost. Let us return to the *Catwalk Closet* example. Although their motivation is neutral (a fancier wardrobe) or even praiseworthy (a wish to reduce one's ecological footprint), customers of the company unknowingly ended up contributing to health ailments and exploitative wages. Questioning and raising awareness of the problematic practices at Catwalk Closet should not be psychologically threatening to the other constituents (of the unorganised collective of customers of the company), although they will presumably not be happy to hear about it either. While their SSO is that they are all customers of Catwalk Closet that turned out to exploit their workers, they were all kept in the dark about the shadier aspects of the company.

At other times, the practice being questioned can feel more personal, like the norms around what food we should eat. In such cases the stakes are higher and there can be more psychological resistance from others with the shared social orientation. Questioning an ingrained habit can feel threatening to others, as it can perceived as judging their behaviour. If you have thought that meat should be eaten daily, but begin to question your earlier beliefs and habits, others who still have that SSO can become defensive if you bring this up with them. It is often too easy to come across as preachy when you have changed something about your own habits and want to tell others about it. It has been suggested that how you communicate to others about your choices matters. If you present the changes you have made yourself as a result of an ongoing process, or focus on specific acts and not the overall character of someone, others are more likely react in a less defensive manner and be more open to change how they act (Brouwer et al. 2022).

Cristina Bicchieri (2017) separates independent collective behaviours from interdependent ones. We act according to our habits, social customs and moral imperatives regardless of what others do, making these collective behaviours independent from the behaviour of others. In contrast, social norms, along with conventions such as signalling systems and fashions, are interdependent: they are undertaken due to our caring about the opinions of others and what they do. If people perceive a norm changing or being challenged, the resulting adjustments in expectations may generate modifications in behaviour (Andrighetto and Vriens 2023). Because daily meat eating usually has a strong social aspect, such as family meals, shared barbeques with friends, and so on, it is a social norm and to change it, we have to question both the empirical and normative assumptions upholding it (and believe that others also find this important). This could be done by starting to offer meat-free options at gatherings with friends and family, to give just one example.

Becoming aware of our shared social orientations is to have our antennas out for the collective context and any changes to it, including staying alert to options to contribute to a debate on values and actions from one's social positions. Recall Zheng's (2021) aspirational morality from Chap. 1. She thinks that while expecting us to be moral angels is wholly implausible, it is reasonable to expect us to strive to take at least some opportunities to pursue moral ideals and contribute to social change. I agree. Perhaps through becoming aware of their SSO, people begin discussions and debates with others in the unorganised collective. Such collectives might even start to view themselves as a social group, positioned within wider societal structures, facing the same kinds of questions. We can surprise ourselves by becoming one of the instigators for collective changes. Still, I think in most cases it is probably too much to *demand* that an individual gets the collective ball rolling (or gathers enough others to get the ball rolling together). It is great if they do, but this is up to them in most cases and depends on their particular situation and characteristics. However, what *can* be demanded is that one should try to have their antennas, their feelers out for possible changes in the collective context. This is not a passive stance, when you are simply required to have an open-minded attitude or a disposition to react when you pick up clues that things are changing. Rather, the concept of shared responsibility I am advocating is an active one, where you intermittently scan your environment for possible changes, in addition to being ready to react if you pick up such signals.

While social norms can be for or against mitigation action, they are not fixed. We humans are social beings who are used to scanning our environments for signals. People look for social cues when they estimate if and when cooperation is socially fitting (Geiger and Brick 2023). If a minority is ready to act despite the prevailing social norms, this can inspire others to join in. Once a certain threshold is exceeded and the social expectations around the issue have changed, the majority begins to follow these new norms, even without changes to their personal normative beliefs

(Andrighetto and Vriens 2023). After all, while there can be many motivations for social change, such as feeling hopeful (Greenaway et al. 2016), wanting to belong is one of our core social motives (Geiger and Brick 2023). Social norms are one of the central predictors of climate mitigation behaviours also (Brick et al. 2021). Yet social norms pertaining to overconsumption are rarely challenged in climate policies, and very few net zero strategies focus on the potentially rapid mitigation impact that behavioural changes to lifestyles could bring (Akenji et al. 2021).

Sometimes it is possible to opt out of a harmful SSO. If ethical options are available and we can afford these, we can choose to buy a more expensive chair from a company that produces its furniture ethically over a chair from a manufacturer with green-washed ethical credentials. However, choosing to opt out and change our SSO assumes that we know about the harmful practices, but this knowledge is not always available. Although the customers of Catwalk Closet contribute to exploitative working conditions at the warehouses with each borrowed piece of clothing, they do not necessarily know anything about them. Being oblivious of harm insulates you from blame, as long as you could not have been reasonably expected to find out. When there is a collective agent between the individual contributor and the collectively caused harm, the collective agent bears a large chunk of the responsibility, perhaps even all of it. The unorganised collective (customers of Catwalk Closet) is centred around an organised collective agent (Catwalk Closet Inc.). The latter is calling the shots when it comes to the salaries and working conditions of its employees. It is working hard to control the image it presents to its customers and to conceal what happens in its dispatching warehouses. You are not blameworthy as a customer of a company that has meticulously hidden its problems from the public eye and that has misleadingly advertised its ethical credentials. In cases like this, the deceiving collective agent could bear all of the responsibility, although in our interconnected world, such situations might be more limited than many of us would like to think. While the organised misinformation campaigns around climate change described in Chap. 2 do complicate the picture for individuals who live in communities in which climate misinformation is abundant, it would be too much to claim that the climate change problem is hidden from us.

Individuals need a reason to act (or refrain from acting) that connects appropriately to the benefit or harm that has been caused collectively. Nefsky (2015) raises a worry that when we employ a broader conception of contributing based on character-based reasons, we can no longer explain why we should take one specific action rather than another. If my action does not necessarily make a difference to the outcome, but is meant to express support and solidarity, then would not such a conception of shared responsibility demand only symbolic acts of us? In responding to such a worry, I think the key is the narrative that the small contributions tell, the pattern that they form. Think of someone who says that caring for the suffering of distant others is a part of their character. They might regularly wear tee shirts by Amnesty International and Médecins Sans Frontières, but this means nothing if they are not also willing to take action sometimes, at least when an opportunity or a clear need arises. To put this in another way, for their choice of clothing to mean something and count as a small contribution in the SSO way, they should be alert to opportunities to make a difference within the unorganised collective that they belong to and act on these at least occasionally. This could entail joining a march, signing a petition, donating money to an urgent appeal, correcting a relative's faulty argument about refugees, or purchasing a new pair of shoes manufactured in a workers' cooperative in a country that usually exports goods manufactured in sweatshops. If they never take up such opportunities, they open themselves up to the charge that they do not actually care, but only want to make it seem that they do. Especially in cases where a direct small choice or a direct small effect is missing, we might rarely have an opportunity to walk the walk. When we do, we should take it. Otherwise displaying all the symbols in the world will not tell anything of substance about us.

The same goes for someone who does nothing to combat the problem of climate change apart from putting a link on their social media profile to a climate activist organisation. Say that this person also works in the PR industry, in a company with links to the fossil fuel industry. A colleague from another firm tells them about the Clean Creatives campaign, but they do not sign up. At the same time, they keep objecting to their group of friends who would want to try a vegetarian restaurant for lunch. They also buy an SUV despite knowing about the environmental impact of such cars, and display behaviours that underline their refusal to make any amendments to their consuming habits in light of climate change. The symbolic act of displaying support for climate action on their social media account is nothing but hypocrisy in such a context. Although small contributions to collective-caused outcomes can sometimes be expressive or symbolic only (this is usually not known at the time of choosing to make such a contribution), we should stay alert to more substantive opportunities to contribute to collective solutions within the unorganised collective.

Constituents of unorganised collectives can become aware of their SSO through their own reflection, or after others point this out to them. This can be done from outside or inside the unorganised collective. I will give two examples of this from both ends of the unorganised collective spectrum in terms of how much there are existing social relations between the constituents. The first is about an interconnected unorganised collective that encompasses members from many different collective agents: the SSO is centred around the constituents' chosen profession and its impacts on the climate. The second example is of less interconnected but nevertheless interdependent unorganised collective of people who influence each other's consumer decisions.

Recall the Clean Creatives campaign example from the previous chapter, and how it is run by people who are working in the field of advertising and PR, making it a form of peer-to-peer campaigning. The campaign targets professionals who are members of various PR and advertising agencies, but it targets them not only as members of particular collective agents, but also more widely as people who work in ad and PR. By appealing to the collective level impacts of their work for the fossil fuel industry and the shared professional values held by them, the campaign appeals to the constituents' social positions as PR and ad industry experts. It does this partly through utilising existing relations, however superficial. Here again the key is becoming aware of the SSO and its impacts. Commenting on the increased self-awareness of creatives working to greenwash the fossil fuel industry, Melissa Aronczyk, an academic who has researched the industry, notes that: "I don't think, historically, PR people saw themselves as being responsible for values the clients they were working with were promising" (quoted in Mufson 2022).[2]

[2] She adds, "It is only when they themselves are dragged into the limelight that they have to respond" (quoted in Mufson 2022). Even when put on the spot, it is all too easy for us humans to tell comforting narratives to ourselves and to others about why we did something. When interviewed by *The Washington Post* about the work his firm Edelman has done for the oil and gas industry over the years, Richard Edelman, president and chief executive of the PR giant said, "I feel wholly comfortable with what we did," adding, "it was 2014. Fracking and all this was quite new. We made the case of that being a positive for America" (Mufson 2022). I find that 2014 is very late in the day to be still allowing oneself to put blinkers on in terms of the effects of the industry on the climate.

Pressure on the unorganised collective can also be exerted from the outside, like when over 450 scientists called for PR and advertising companies to cut ties with the fossil fuel industry, asserting that the campaigns they run "represent one of the biggest barriers to the government action science shows is necessary to mitigate the ongoing climate emergency" (Mufson 2022). Again, such campaigns target both the collective agents and their members, but in a wider context that just as members of some particular collective. Rather, the campaign focuses on social positions and targets all creatives who work to greenwash the fossil fuel industry, through pointing to their SSO and the impact it can have. The discussion within and outside the profession on what should be the standards regarding working with fossil fuel companies has arguably increased the awareness of the creatives and other professionals working in advertisement and PR regarding their role in the bigger picture. This has prompted some to take action and to get organised to work together to bring about change. While there is no umbrella organisation to green up the PR and advertising industries, there are now smaller active organised groups within the larger unorganised group of professionals, trying to get everyone to change their thinking around acceptable business practices.

The second example is of an unorganised collective of consumers without much interconnectedness between the constituents in terms of relations. Consumer demand for SUVs has been growing exponentially in recent times and the global fleet of SUVs has ballooned from less than 50 million in 2010 to around 320 million in 2021 (Cozzi and Petropoulos 2021). In the case of SUV owners, their SSO is valuing things like car ownership and private transportation and the look or versatility of SUVs. It is also about the indirect interaction of being influenced by what cars others are driving and in turn influencing the decision of others in the streets and in your neighbourhood (SUVs tend to be popular in the better-off residential areas). Consumer behaviour can also be influenced by the mistaken idea that SUVs are safer, although they do not perform better for passenger safety, but do put other road users in higher danger (Logan and Newstead 2018). Thus, the increased number of SUVs on roads can put pressure on other drivers to start driving bigger cars also, due to normal-sized cars being at a disadvantage when in crashes with SUVs. This means that there is a certain interdependence between private decisions and what other drivers choose to do. However, consumers might not be aware of the problematic nature that their individual choices have in aggregate, or how their choices are interdependent on the existing infrastructure and what others do (see Chap. 1). Although the examples such cases offer to responsibility discussion are not as clear as those offered by the more interconnected unorganised collectives are, they are worth discussing also.

Such lifestyle choices, such as your primary mode of transport, affect our overall emissions to a much greater extent than one-off consuming decisions. Because SUVs are heavier than a medium-sized car, they consume around 20% more energy compared to them. This translates to SUVs ranking among the top causes of growth in energy-related CO_2 emissions over the past decade, with over 35 million SUVs being bought and sold in 2021 alone, driving up annual emissions by 120 million tonnes towards the over 900 million tonnes of CO_2 emitted by SUVs in total in 2021 (Cozzi and Petropoulos 2021). Once such figures are brought up more and questioned by the consumers, trends might change. This does not shut out the need for stricter regulation, but simply points out the complex relation between supply-demand and the way consumer trends can drive up emissions. This should not be forgotten in responsibility debates, especially as legislation often takes time. When changes in behaviour originate from the consumers themselves, such action can also face less resistance than restrictions imposed from the outside. SUVs are popular in electric vehicles also, but the larger batteries required in SUVs can undermine the environmental benefits of such electric vehicles (Irfan 2022). Those consumers with a predilection for both large cars and environmentally friendly alternatives might come to question their choice once they become aware of the wider picture their choices take place in. A driver who cares about climate change might change their mind about buying an SUV, electric or otherwise, and opt for another model instead, thereby opting out of the set of drivers who are pushing up emissions through their choice of a car model. The actions of the constituents of unorganised collectives without much interconnectedness can give signals and incentives to collective agents, plus it can feel empowering for the constituents if they are able to opt out.

Awareness of SSO is only the potential starting point for change, but it can facilitate reassessment of the collective patterns of behaviour. What I find to be the key, is that when we pay attention to our surroundings and our own social positions in the world, we can start noticing harmful patterns that we are enmeshed in (and/or impacted by) or notice what needs changing and where we might be able to help. SSOs can also be about a vulnerable position one shares with others, so it is not just about harms one is implicated in one way or another. However, as the title of my book refers to taking responsibility for climate change, I will be focusing on the SSOs in which we are or become aware of our connections to something harmful, or of our potential to do something good together. I should note, however, that I do not suggest that there is an automated link between knowing about a harm and our connections to it—or our potential to do something good together—and changes in our behaviour. People react to new information differently and process it from their own unique viewpoint. They also respond differently to normative claims. Not only is there heterogeneity between individuals in terms of how sensitive they are to norms and risks, but the strength of social norms also varies across societies (Andrighetto and Vriens 2023). Some matters will be much more contested than others will and some unorganised collectives will be more open to exploring their SSO than others. It matters a lot who the messenger is as well, and if they are seen as being in-group or out-group. Awareness of interconnectedness increases with debate and is only ever the starting point for action or (re)articulating of shared values.

Shared social orientation evolves around interdependent activity, meaning that the actions and opinions of others can affect the behaviour of a constituent. The more professionals join the Clean Creatives campaign, the more pressure it puts on others to follow suit. It also becomes less costly to join action where many others have already joined in. As Judith Lichtenberg (2014) has observed, moral demandingness is not a fixed quantity, but dependent on what other people around them do. Becoming aware of one's SSOs should matter to the actions of individuals: it allows us to have a richer and more realistic view of ourselves as moral agents in the interdependent world we live in than the more atomistic view. Individuals are constituted relationally, meaning that our actions have moral significance both in the local and wider contexts (Hourdequin 2010). For example, someone's choice to not to eat meat can be trailblazing in their local community and inspire others to give vegetarian alternatives a go as the option becomes normalised through everyday interactions (eating lunch together etc.). At the same time, it is a small contribution to the wider trend of increasing customer demand for vegetarian food.

What tangible action a shared social orientation calls for is a question that cannot be answered independently of the context, including the position and characteristics of a given constituent. One factor is the extent of social relations between the constituents. Pre-existing linkages, however superficial, will make it easier to speak to your fellow constituents. They also make it easier to organise collectively, be it the real-life example of the

PR and communications industry professionals targeted by the Clean Creatives campaign, or the invented and looser unorganised collective of the Catwalk Closet customers. At the other end of the spectrum, we have people picked up only by their consuming habits, like SUV drivers or daily consumers of meat, where the existing personal linkages between the constituents could be minimal or non-existent.

Despite these differences, there is a systemic background to their behaviour that enables it (e.g., factory farming, meat-focused cookbooks, the kind of food available in the school canteen, etc.), with interdependent activity. Even at this loosest end of SSO, there are social aspects to the activities of individuals: constituents uphold these structures through their contributions for demand of cheap meat, and through their support for social standards that normalise and encourage such lifestyle choices.[3] In these cases, the SSO might call for opting out of the unorganised collective. Note, though, that there might be various smaller mini collectives within the larger unorganised collective, like family units or flatmates, or groups of co-workers that eat lunch together regularly. Within these small collectives, individuals can have strong existing social relations and potentially a lot of influence over the choices and thinking of others. Consumption choices can also act as signals to others as to what norms we are willing to support (when we have the everyday luxury to choose). It is probably within such smaller overlapping groups that change in norms starts taking place and new habits are formed.

Consumers as Constituents

When there is a lot of interconnectedness between the constituents in terms of existing relations, like in the Clean Creatives campaign, it is easier to see how SSO might motivate us to act than when we are discussing very loose unorganised collectives, such as SUV drivers. Despite the inevitable vagueness in moral arguments at this looser end of the spectrum, I want to take a closer look at consumers in this section. I discussed in Chap. 2 how the fossil fuel industry has been deliberatively pushing for a

[3] The United Nations Environment Programme suggests that food systems, which account for one third of all emissions, can be reformed to deliver rapid cuts in emissions, and include the lifestyle consuming patterns of individual citizens as one of the methods for this, along with action by national and local governments, cities, and the private sector (UNEP 2022).

misleading framing focusing on individual carbon footprints, diverting attention away from the changes required in energy infrastructure. Here I want to argue that despite this misleading framing, how consumers respond to climate change is still relevant. After all, supply and demand is not a straightforward issue or a one-way street. However, any mitigation responsibility we might have as consumers is nearly always tempered by the structures that our consuming choices take part in (see Chap. 1). Thereby individual consumption should not be discussed independently of the structures that shape our consumption.[4] I suggest that the SSO framing can help us to appreciate the structures and social norms that need chancing.

Most emissions are driven by routine consumption. GHG emissions can be categorised under five sectors of economic activity: energy supply systems, industry, AFOLU (agriculture, forestry, and other land use), transport, and buildings (IPCC 2022). A significant percentage of all these emissions come from everyday consumption choices: where we live, what we eat, how much we move around, and by which mode of transport. Intuitively, there seems to be differences between the permissibility of emissions in terms of the consumption categories they fall under. Luxury emissions come from consumption that one could forgo, like buying the latest phone model or yet another pair of shoes, whereas subsistence or survival emissions are from consumption required to satisfy basic human needs (Agarwal and Narain 1991; Shue 1993). Even when it comes to the consumption we need, some consumer choices are more frivolous than others. It might be that you need a car to get to work, but you are not happy with just any car, but want a big four-wheel-drive car, for example. Still, there is arguably tension with all consumption, frivolous or not, and the goal of reaching net zero by 2050 (Duus-Otterström 2023).

When discussing consuming decisions, it is important to keep in mind that many of these choices are complicated and influenced by many factors. Different things will have different meanings to different people and groups. For this reason, consuming discussions can feel intensely personal, and people might (rightfully) feel resentful if they are told what to do, or something is branded outright as the “wrong” choice. In such situations, our defence-mechanisms can easily kick in. When people are confronted with fellow consumers who have adopted new ways out of moral concerns, they can become defensive if they have failed to adopt it themselves, as this

[4] Individual direct climate responsibility will be the topic of the next chapter.

can threaten their moral self-conception (Bolderdijk et al. 2018). Although I argued earlier that some feeling of discomfort can be beneficial in order to change our ways, this should not be overdone and is probably best if it comes from within. Having said that, it is possible that although the initial reaction is defensive dismissal, in the long term, others might be motivated to follow their example, especially if the choices of the innovators are based on moral concerns (Bolderdijk et al. 2018).

Furthermore, our preferences are not set in stone, but can change with the times, exposure, practice etc. It is not surprising if someone who has eaten meat as part of every meal since they were a child will view vegetarian food with suspicion at first. But when they are exposed to the array of flavours and textures available (e.g., in restaurants that offer skilfully prepared vegetarian meals), their preferences are likely to change. Maybe they will learn to use spices to flavour home-cooked vegetarian food to their taste. It will not then feel like something has been taken from them when they reduce meat consumption, but rather that something new and exciting is introduced to them. They might even choose to go fully vegetarian without any yearning for meat. The same can be said for most consuming choices: there are opportunities when we amend our ingrained habits. Bicycling or walking instead of driving can benefit us through increased exercise, living in a less expansive home translates to less time spent cleaning, having a more selective take on buying clothes can help us to have a wardrobe that we actually use, and so on. Change can be as much about gains as it is about giving up something.

Available technology also affects the choices of individuals. Perhaps people will soon not have to make choices based on flavour and texture, as protein production is changing radically by new innovations.[5] This is at least the story if you believe what those who analyse and forecast technology-driven disruptions are saying. In a report by RethinkX, an independent think tank, the cost of proteins is projected to become many times cheaper than current animal-derived proteins, so the replacement will be primarily driven by economics (Tubb and Seba 2019). Cow's milk and meat are anticipated to be the next proteins the production of which

[5] Not only cheaper, the new modern proteins are expected to also have better taste and nutritional value than animal-based proteins. The new technology will utilise innovations in precision biology, including engineering molecules that will produce proteins through fermentation in decentralised, localised farms close to towns and cities. This will result in significant reduction in GHG emissions from food sector, as well as land currently used for livestock and feed production being freed up for other uses (Tubb and Seba 2019).

will be disrupted by precision fermentation in the coming years. The report identifies consumers as one of the four main groups that could accelerate or slow down the oncoming food disruption—along with businesses, investors, and policymakers—recognising the interdependent nature of these groups, where actions and choices of one affect also the other groups. After all, supply and demand is a two-way street, both driving the other.

Collective agents can offer incentives that shape individuals' consumption patterns. For example, utility companies can offer lower energy prices to their customers to run their appliances during off-peak hours, reducing demand spikes for electricity production (Irfan 2022). But collective agents can affect consuming habits through many methods, such as signalling with prices, nudging individuals through choice architecture (like when a restaurant offers the vegetarian options first on its buffet or highlights them on its menu), making their products greener, innovating new climate friendly options, and so on. States and governments have the most powerful tool for this: legislation. I will discuss the possibility of curbing certain options in the next chapter. But in this chapter my argument has been that what consumers decide to do matters as well. Moreover, it can feel empowering to be part of the solution and to drive change forward.

When existing consumer norms and habits are challenged from the grassroots, by the consumers themselves, it can give powerful incentives to collective agents in terms of consumer demand or support for greener policies, plus feel empowering for the individuals. After all, consumers can also be powerful, like when their combined conservation actions reduce demand on energy grids during peak demand periods (Irfan 2022).[6] Empirical research also suggests that amongst consumers, there might already be a perception of a shared responsibility for climate change action (Wells et al. 2011). In any case, I believe it is false to draw a sharp line between consumption decisions based on awareness of SSO and very low-cost political action (such as signing petitions or signalling your support for political causes by wearing badges or tee shirts), at least in those societies that are currently hyper-consumerists. Furthermore, consumers employing their power through measures such as boycotts within

[6] A heat wave in California in September 2022 made the electricity demand skyrocket, straining the state power grid, prompting California's Office of Emergency Services to send out a mobile phone alert to conserve energy across targeted counties to prevent rolling blackouts (Irfan 2022). It worked: demand dropped dramatically within minutes.

well-functioning market democratic societies can be a democratically legitimate means of promoting justice (Berkey 2021). None of this is to try to deny that straight-up political action is often the more robust way for an individual to try to make a difference.[7]

Of course, having a choice in consumption habits to begin with is a luxury not afforded to everyone (overstuffed closets and decisions about what to eat when options are abundant are very much first world problems). Take as an example a poor family who owns a goat that provides them with milk and eventually meat, helping them to meet their protein needs. There is a world of difference between this scenario and someone who could satisfy their protein needs in other ways. Choosing to eat meat at every available opportunity despite having other options amounts to easily avoidable luxury emissions. Vast inequalities in choice also exists within nations (see the next chapter), so it can also be a tricky terrain to make anything resembling a judgement on someone else's eating habits (or other consumption) if you are from different economic classes. It can be just as jarring as the wealthy preaching to people on low incomes to eat

[7] The usual solution offered to worries over harms done together and the potential inconsequentialism of individual acts is that we are collectively obligated to try to bring about reductions in GHG emissions through coordinating our efforts in the political arena (e.g., Cripps 2013; Johnson 2003; Sandberg 2011). This could be through voting decisions, taking part in demonstrations, signing petitions, beginning your own campaign, or taking any other political action in support of mitigation policies. For example, Elizabeth Cripps (2013, 148) argues that although individual political actions might not "succeed straight off, they can still contribute to a stockpile of impetus for collective change." She likens taking political action to putting money in a bank: political actions are not "throw-away acts", unlike unilateral individual emissions reductions might be. However, it has been argued that this begs the question, as the effectiveness of participating at demonstrations can be called into question in the same way that making changes to one's consumption habits can (Peeters et al. 2015, 86–87). Noting how the impact that an average individual can anticipate in getting their government to act is very small, Aaron Maltais (2013) suggests that any individual obligation to take political action for climate change mitigation must also consider the low likelihood of effective collective action. According to him, what can be demanded of an average individual is not much: any duties must be low-cost, such as voting for a green candidate, because high personal costs will most likely be wasted and therefore cannot realistically be demanded from us. Still, even if appealing to what we should do as citizens might not offer a straightforward solution to worries around inconsequentialism, this does not mean that we should not take our responsibilities as citizens seriously. I discuss the potential of participatory democracy in Chap. 6. Here I only want to note that although my focus in this book is on the areas of climate change responsibility that I think do not get enough attention, I do not wish to deny the importance of paradigm political action, such as voting, protesting, or lobbying and petitioning the state. These are all very important and are discussed widely in the literature.

less junk food and more of the green stuff, even though the local grocery shops may not even stock anything but processed foods in the more deprived areas of cities. That is why questioning consumption should be done mostly within the luxury emissions side of things. Still, differentiating what is essential and what is easily expendable consumption cannot be done in broad categories only since this will vary from one person to the next. For some, clothes are an important part of expressing their identity, while for others snowboarding is what makes them tick. For many others, these are just fun things, easily replaceable by other goods if needed. Non-judgemental debates within SSO can help to identify potential for changes.

It is thus hard and perhaps even misguided to make blanket statements about the frivolousness of the choices of others because there is a lot of variation from one person to the next on what things matter to them and what gives them meaning. Therefore, it is best to start with oneself, although you should be willing to prioritise, innovate, and compromise. In some cases, this means giving a fair shot to trying to change your preferences. Saying that, there are some exceptions to this overall narrative, namely certain high-end consumption choices that simply have such heavy carbon footprints (like super yachts and space travel) that they should be abstained from and/or regulated against, at least until technology makes them safe choices for the planet. I will discuss these in the next chapter. What usually works the best is when someone questions their own habits and is willing to engage others with the shared social orientation. For example, an enthusiastic golfer could try to get their golf course to become more sustainable through measures such as capturing rainwater for irrigation or allowing the courses to dry out when there is not enough water.

Again, change is harder in some cases than others. Taking climate mitigation seriously if you are a dairy farmer or a meat producer is a much taller order than for the golf enthusiast, of course. But even then, it is possible; people can transition their livelihoods. For example, in Scotland there is a grassroots organisation called Stockfree Farming (formerly known as Farmers For Stock-Free Farming), which specialises in supporting meat and dairy farmers who want to transition to animal-free agriculture.[8] Its staff includes people who had previously been livestock farmers or worked at such farms. Another UK organisation, Refarm'd, helps dairy farmers to make the switch to producing plant-based milk, cheese and yoghurt products. These are organisations build around SSO. Former

[8] See stockfreefarming.org/about-stock-free-farming.

dairy farmers have swapped from cows to farming oats, but such changes are not always easy (Bearne 2022). Aside from practical challenges in the transitioning away from animal-based farming, farmers who are seen to "switch sides" can face the ire from their neighbours and fellow farmers, so how others react impacts the social costs to the individual in these kinds of cases, too.

The responsibility qua constituents of unorganised collectives should not be conceptualised as that of isolated individuals emitting in a social vacuum, but instead through the various structures and patterns that such emissions are part of in the network of interdependencies that connect us. The responsibility of those who participate in the practices and norms that upheld systemic harms is to evaluate them critically together. This can mean engaging in political action by creating public debate or by supporting and envisioning political action around a SSO. Social movements are born when constituents of unorganised groups start coming together. We should together seek ways to assess and question the practices, norms, and institutions that have created climate change and to try to seek solutions to address and mitigate the harms. Sometimes this will mean creating new collective agents (e.g., Cripps 2013; Gardiner 2011).

Concluding Remarks

Although acting as members of organised collectives is often the main way in which we can hope to make a difference in collective action problems, we should not feel indifferent towards being a constituent of some normatively significant unorganised collective. All such interdependencies affect the coherence of our moral lives. Not every choice in every single instance is meaningful, but we can look at the patterns they form, and it is the bigger picture that matters; what are we regularly involved in, what are we willing to be associated with, what do we tolerate, where are our limits, do we take any action to try to remedy a situation, and so on.

My argument in this chapter has been that we have a responsibility to stay alert to our shared social orientations as much as is feasible. There might be no specific action that can be pinned on one due to our SSO that links us to a collective outcome, but we should stay alert to occasions to act and be part of change when collective opportunities arise. Some individuals are better positioned than others to effect change; so what you should do depends partly on the social positions and roles you occupy, along with your personal characteristics and resources.

By advocating the responsibility to be alert to such opportunities, I do not mean to suggest that we should feel the weight of the world on our shoulders or that we should try to fix everything. Large-scale structural harms are always systemic, and an individual cannot change things unless enough other people are willing to join them. It is easy to feel powerless in the face of such obstacles. As Allison Briscoe-Smith (2020) observes, solutions to human-made ills, such as systemic police brutality to black Americans, can "feel out of reach to the average person", and this can add to the stress and fear felt by an individual. We should not feel that we failed if we tried to be part of a solution and a systemic change did not come about. Instead, we should feel proud that we tried. That said, social change can sometimes happen fast, especially in unusual times. Times like these can present opportunities for individuals to get started on change even in systemic issues. At the minimum, we should not think our small contribution to collective outcomes is necessarily morally neutral just because the unstructured collective is large, especially when the outcome is morally significant. Our shared social orientation can guide us to see the collective larger picture of what it is that we are involved in and what shared values these actions express.

References

Agarwal, Anil, and Sunita Narain. 1991. *Global Warming in an Unequal World: A Case of Environmental Colonialism*. New Delhi: Centre for Science and Environment.

Akenji, Lewis, Magnus Bengtsson, Viivi Toivio, Michael Lettenmeier, Tina Fawcett, Yael Parag, Yamina Saheb, Anna Coote, Joachim H. Spangenberg, Stuart Capstick, Tim Gore, Luca Coscieme, Mathis Wackernagel, and Dario Kenner. 2021. *1.5-Degree Lifestyles: Towards A Fair Consumption Space for All*. Berlin: Hot or Cool Institute.

Andrighetto, Giulia, and Eva Vriens. 2023. A Research Agenda for the Study of Social Norm Change. *Philosophical Transactions of the Royal Society A* 380: 20200411.

Ásta. 2018. Social Kinds. In *The Routledge Handbook of Collective Intentionality*, ed. Marija Jankovic and Kirk Ludwig. Routledge.

Audi, Robert, and Patrick E. Murphy. 2006. The Many Faces of Integrity. *Business Ethics Quarterly* 16: 3–21.

Bearne, Suzanne. 2022. The Meat and Dairy Farmers Who Are Going Vegan. *BBC News*, May 26. www.bbc.com/news/business-61565233.

Berkey, Brian. 2021. Ethical Consumerism, Democratic Values, and Justice. *Philosophy & Public Affairs* 49 (3): 237–274.

Bicchieri, Cristina. 2017. *Norms in the Wild: How to Diagnose, Measure, and Change Social Norms*. New York: Oxford University Press.

Bolderdijk, Jan Willem, Claire Brouwer, and Gert Cornelissen. 2018. When Do Morally Motivated Innovators Elicit Inspiration Instead of Irritation? *Frontiers in Psychology* 8: 2362.

Brick, Cameron, Anna Bosshard, and Lorraine Whitmarsh. 2021. Motivation and Climate Change: A Review. *Current Opinion in Psychology* 42: 82–88.

Briscoe-Smith, Allison. 2020. How Can I Stay Positive for My Kids When I'm So Overwhelmed? *Greater Good Magazine*, June 4. Berkeley: The Greater Good Science Center at the University of California, Berkeley.

Brouwer, Claire, Jan Willem Bolderdijk, Gert Cornelissen, and Tim Kurz. 2022. Communication Strategies for Moral Rebels: How to Talk About Change in Order to Inspire Self-Efficacy in Others. *WIREs Climate Change* 13 (5): e781.

Cozzi, Laura, and Apostolos Petropoulos. 2021. *Global SUV Sales Set Another Record in 2021, Setting Back Efforts to Reduce Emissions*. Paris: IEA. www.iea.org/commentaries/global-suv-sales-set-another-record-in-2021-setting-back-efforts-to-reduce-emissions.

Cripps, Elizabeth. 2013. *Climate Change and the Moral Agent: Individual Duties in an Interdependent World*. Oxford: Oxford University Press.

Duus-Otterström, Göran. 2023. Subsistence Emissions and Climate Justice. *British Journal of Political Science* 53: 919–933.

Gardiner, Stephen M. 2011. *A Perfect Moral Storm: The Ethical Tragedy of Climate Change*. New York: Oxford University Press.

Geiger, Nathaniel, and Cameron Brick. 2023. Core Social Motives Explain Responses to Collective Action Issues. *Social and Personality Psychology Compass* 17 (3): e12732.

Greenaway, Katharine H., Aleksandra Cichocka, Ruth van Veelen, Tiina Likki, and Nyla R. Branscombe. 2016. Feeling Hopeful Inspires Support for Social Change. *Political Psychology* 37 (1): 89–107.

Held, Virginia. 1970. Can a Random Collection of Individuals be Morally Responsible. *The Journal of Philosophy* 67: 471–481.

Hormio, Säde. 2017. *Marginal Participation, Complicity, and Agnotology: What Climate Change Can Teach Us About Individual and Collective Responsibility*. Helsinki: University of Helsinki.

———. 2023. Collective responsibility for climate change. *Wiley Interdisciplinary Reviews Climate Change* 14 (4): e830.

Hourdequin, Marion. 2010. Climate, Collective Action and Individual Ethical Obligations. *Environmental Values* 19: 443–464.

IPCC. 2022. Summary for Policymakers. In *Climate Change 2022: Mitigation of Climate Change. Contribution of Working Group III to the Sixth Assessment Report of the Intergovernmental Panel on Climate Change*, ed. P.R. Shukla, J. Skea, R. Slade, A. Al Khourdajie, R. van Diemen, D. McCollum, M. Pathak, S. Some, P. Vyas, R. Fradera, M. Belkacemi, A. Hasija, G. Lisboa, S. Luz, and J. Malley. Cambridge: Cambridge University Press.

Irfan, Umair. 2022. Consumers, Not Corporations, Saved the Power Grid. What Else Can We Do? *Vox*, October 31.
Johnson, Baylor L. 2003. Ethical Obligations in a Tragedy of the Commons. *Environmental Values* 12: 271–287.
Kutz, Christopher. 2000. *Complicity: Ethics and Law for a Collective Age*. Cambridge: Cambridge University Press.
Lichtenberg, Judith. 2014. *Distant Strangers: Ethics, Psychology, and Global Poverty*. New York: Cambridge University Press.
Logan, David, and Stuart Newstead. 2018. I've Always Wondered: Are SUVs and 4WDs Safer Than Other Cars? *The Conversation*, September 23.
Maltais, Aaron. 2013. Radically Non-Ideal Climate Politics and the Obligation to at Least Vote Green. *Environmental Values* 22: 589–608.
May, Larry. 1992. *Sharing Responsibility*. Chicago: The University of Chicago Press.
Mufson, Steven. 2022. More Than 450 Scientists Call on PR and Ad Firms to Cut Their Ties with Fossil Fuel Clients. *The Washington Post*, January 19.
Nefsky, Julia. 2015. Fairness, Participation, and the Real Problem of Collective Harm. In *Oxford Studies in Normative Ethics*, ed. Mark Timmons, vol. 5, 245–271. Oxford: Oxford University Press.
Peeters, Wouter, Andries De Smet, Lisa Diependaele, and Sigrid Sterckx. 2015. *Climate Change and Individual Responsibility: Agency, Moral Disengagement and the Motivational Gap*. Basingstoke: Palgrave Pivot, Palgrave Macmillan.
Sandberg, Joakim. 2011. 'My Emissions Make No Difference': Climate Change and the Argument from Inconsequentialism. *Environmental Ethics* 33: 229–248.
Shue, Henry. 1993. Subsistence Emissions and Luxury Emissions. *Law & Policy* 15 (1): 39–60.
Tubb, Catherine, and Tony Seba. 2019. *Rethinking Food and Agriculture 2020-2030: The Second Domestication of Plants and Animals, the Disruption of the Cow, and the Collapse of Industrial Livestock Farming*. A RethinkX Sector Disruption Report, September 2019.
UNEP. 2022. *Emissions Gap Report 2022: The Closing Window — Climate crisis calls for rapid transformation of societies*. Nairobi: United Nations Environment Programme. www.unep.org/emissions-gap-report-2022.
Wells, Victoria K., Cerys A. Ponting, and Ken Peattie. 2011. Behaviour and Climate Change: Consumer Perceptions of Responsibility. *Journal of Marketing Management* 27 (7–8): 808–833.
Williams, Bernard. 1981. *Moral Luck: Philosophical Papers 1973—1980*. Cambridge: Cambridge University Press.
Young, Iris Marion. 2011. *Responsibility for Justice*. New York: Oxford University Press.
Zheng, Robin. 2021. Moral Criticism and Structural Injustice. *Mind* 130 (518): 503–535.

CHAPTER 5

Carbon Inequality and Direct Responsibility

Introduction

The amount of greenhouse gases (GHGs) that can still be emitted to the atmosphere is very limited if global total emissions are to stay below dangerous levels, as discussed in Chap. 1. However, how to divide the world's carbon budget in a fair manner is a question that remains unresolved. This applies to both theoretical and practical debates around what emission-reduction targets states should have, how much they should be allowed to emit, and for how long. Another unresolved issue is what emissions should count in the budget of a state: emissions that take place within its borders (production-based emissions) or also emissions that happen in other countries but have been enabled by that country, such as through extraction of raw materials that have then been exported (consumption-based emissions). What makes this especially difficult is that emissions are very unequal between regions and states, both historically and at present. These differences are notably stark at the per capita level, i.e., when the size of the country's population is taken into account.

I will discuss carbon inequality in this chapter and what this means for individual direct climate change responsibility. I want to focus on the vast differences between individuals *within* national carbon budgets, regardless of the form such budgets would take. Simon Caney (2009) has argued that allocating emission permits to states is morally problematic in the first place, as doing so would treat them as homogeneous units, even though

S. Hormio, *Taking Responsibility for Climate Change*,
https://doi.org/10.1007/978-3-031-51753-2_5

there are large differences between citizens in how much they emit. Paul Baer et al. (2010) likewise argue that it is problematic to assume that governments represent the interests of all their citizens in climate negotiations. The real effects on the state's citizens will vary according to their economic class and how equal the state is. Thus, when mitigation policies are designed, the socio-economic implications need to be investigated. This is especially important in countries like India where there is a need to provide basic infrastructure and amenities to the unprivileged classes (Sri and Banerjee 2023).

There are thus two sides to the carbon inequality story: the inequality of citizens in a state (inequality within country) and global inequality of citizens of different countries (inequality between countries). The trend that intercuts these two is what I will call the *global inequality of personal carbon footprints*. In fact, in addition to existing international inequality in carbon emissions, emission inequalities between individuals within countries are greater than ever (Chancel 2022). This, I believe, is an important issue that does not yet get enough attention in climate ethics.[1]

Throughout the previous chapters, I have argued that our shared responsibility for climate change is conceptually prior to direct individual responsibility. While this holds for almost everyone, there is a relatively small group of individuals who form a possible exception to this rule. I will call these people *super-polluters*, as their lifestyles result in emissions many times over the carrying capacity of the Earth. I noted in the previous chapter how our responsibility as consumers is nearly always tempered by the structures that our consuming choices take part in (see also Chap. 1). Therefore individual consumption should not be discussed independently of the structures that shape our consumption. This applies even when we discuss direct responsibility, as this is also enabled and driven by structures. I begin by looking at carbon inequality and how emission patterns are widely inequal, both globally and within states. I draw on data from various sources, so the figures I use are meant to only illustrate the scale of the problem, not to carry the weight of the argument as such. I then

[1] An additional feature that makes focusing on the inequal emissions and their causes timely is the increasing dexterity of methods of tracking emissions. While there is currently nothing but estimation calculators available to track individual carbon footprints, calculating and estimating emissions is getting more sophisticated by the day. One example is Climate Trace (climatetrace.org), which uses satellites and other remote-sensing technologies to spot emissions where they take place and then combines these data with direct measurements and AI to build models to estimate emissions at their source.

introduce the concept of super-polluters. When it comes to climate harms, I will argue that individual direct responsibility is limited to very wealthy individuals with high emissions. This individual direct responsibility is to not increase the probable risk of serious harm—the deprivation of fundamental capabilities—to others, as long as they can do so at a less than significant cost to themselves. I will also argue that offsetting is not a reliable way to meet it. Instead, what should take focus are mega-polluting consumption and the investment choices of the super-polluters.

Emissions Inequality

Global emission patterns are widely unequal. China, the United States, the European Union (EU), India, Indonesia, Brazil, the Russian Federation, and international transport contributed more than 55% of the total global GHG emissions in 2020, while the G20 countries contributed 75% of the total emissions (UNEP 2022, 7). Half of the global population emits 11.5% of the total global emissions, while the top 10% of emitters globally are responsible for 48% of the total emissions (Chancel 2022). It has been estimated that between the first IPCC report in 1990 and the 2015 Paris Agreement, the carbon emissions from the consumption of the world's richest 1% were twice that of the combined emissions of the poorest 50% of the global population (Kartha et al. 2020). The carbon footprints of the top 1% of emitters globally were more than 1000 times greater than those of the bottom 1% of emitters in 2021 (Cozzi et al. 2023). There are vast differences between the per capita emissions of an average North American or Australian and the emissions of an average African or Latin American citizen. Many of these inequalities have their roots in extractive colonial economies and ongoing patterns of uneven exchange (Newell and Adow 2022). Saying that, the rich individuals around the world are often emitting just as much as the rich individuals in the Global North. While many of the top emitters live in North America, other regions with very high per capita emissions in the top 10% group include East Asia, Russia and Central Asia, and the Middle East (Chancel 2022).

In recent decades, the differences in GHG emissions of individuals have decreased at the between-countries level. This is due to the rise of the top- and middle-income groups in developing countries, combined with the relative stagnation of incomes and emissions of the majority of the population in industrialised economies (Chancel and Piketty 2015).[2] The average citizen of a rich country in 1990 had much larger emissions than people in the rest of the world, so the majority (62%) of global carbon inequality used to be because of these between-country differences (Chancel 2022). The situation is nowadays different as emission inequalities within countries has seen an upward trend. While the emissions from low- and middle-income groups within rich countries have declined since 1990 (driven by climate and energy policies and industrial efficiency gains but also compressed wages and consumption), the per capita emissions of the global top 1% have increased, so within-country differences now account for nearly 63% of global emissions inequality (Chancel 2022).

Carbon inequality is more pronounced in some countries than others. For example, in the United Kingdom, the top 1% of earners by income generates roughly the same carbon emissions in a year as the bottom 10% have generated in 26 years (Garcia and Stronge 2022). In China, the per capita footprint of the richest 10% is 33 times the size of the poorest 10%, which is a greater difference than in the United States, where the richest decile emits 16 times as much as the poorest decile (Sengupta 2023). However, this does not mean that overall consumption levels are not very high in the United States. On the contrary, the richest 10% in the United States emits 56.5 tons of carbon dioxide per person per year on average, which is nearly double that of the richest 10% in China and more than double the emissions of the richest 10% in Europe (Sengupta 2023). The emissions of the poorest half of the population in the United States are comparable to what the European middle 40% emit (Chancel 2022). In fact, every EU income group has lower carbon footprints than its US equivalent, partly thanks to less emissions-intensive power grids (Cozzi et al. 2023). This again highlights the importance of available infrastructure when it comes to discussing personal carbon footprints. Furthermore, even though the richest 10% in the EU emits in comparison 'only' around 24 tonnes of carbon monoxide per capita, internal inequalities are similarly large compared to the United States: the top 10% emits around 16 times more than the poorest 10% (Cozzi et al. 2023).

It is important to note that although the global inequality in individual emissions is now mostly due to a gap between low and high emitters within countries, significant inequalities in emissions between countries and regions have not disappeared (Chancel et al. 2023). This means that,

[2] The study looked at data between 1998 and 2013.

in addition to great international inequality in carbon emissions, emission inequalities between individuals within countries are greater than ever (Chancel 2022). This is what I labelled as the global inequality of personal carbon footprints. Vastly inequal carbon footprints within a country are not restricted to just wealthy countries that have high per capita emissions, and low per capita emissions do not automatically translate to only a few very high emitters among the state's citizens. Instead, it seems more likely to find potential high emitters in countries with high economic inequality (e.g., South Africa, United States, Brazil, India), regardless of the per capita emissions of that state. Therefore, we should not forget the other side of carbon inequality: the vast inequalities between individuals in their carbon footprints, regardless of what countries they live in. For example, India has a low per capita emissions level but contains many wealthy people with high carbon footprints. The latter kind of high emitters ought to also pay the appropriate cost for their emissions, regardless of the emissions of their state (Caney 2009).[3]

The global inequality of personal carbon footprints is important for climate ethics because the global top emitters are driving up the emissions. Nearly half of the growth in emissions came from the top 10%, with the top 5% contributing a whopping 37% of the global growth in emissions (Newell et al. 2021). Even more strikingly, the top 0.1% contributed about two-thirds of the entire growth in emissions associated with the poorest half of the global population (Chancel 2022). The total emissions associated with the richest 1% are projected to continue to increase, and it is estimated that by 2030, the richest 1% of the world population will have emissions footprints about 30 times higher than the global per capita level compatible with the 1.5 °C goal (Gore 2021, 3). It should be noted that over 70% of the emissions generated by the global top 1% comes from their investments rather than their consumption (Chancel 2022).

These patterns point to a wider problem for climate action. As Oxfam International and the Institute for European Environmental Policy state in their joint agency briefing note on carbon inequality (Gore 2021): "Maintaining such high carbon footprints among the world's richest people either requires far deeper emissions cuts by the rest of the world's

[3] Caney (2009) suggests that an auction-based scheme of allocating emission permits is more sensitive to inequality within states. This is because the firms who buy emissions rights are likely to pass the costs onto consumers, meaning that those who consume more will bear the costs of their decisions.

population, or it entails global heating in excess of 1.5 °C above pre-industrial levels. There is no other alternative."

As I see it, carbon inequality has two main implications. First, countries with high carbon footprints should do more for meaningful climate action than countries with low footprints. This also means that relatively wealthy individuals within these countries have weightier responsibility as citizens (responsibility as members) to push for meaningful climate action than citizens of countries with low-carbon footprints. Second, people with very high personal carbon footprints, regardless of where they live, have direct personal responsibility regarding taking meaningful climate action. The second implication is what I will focus on in this chapter. While I have used many percentages in this section to illustrate the striking inequalities in emissions that exist at both the global and national levels, what I will concentrate on for the remainder of this chapter are the individuals at the very top of this emissions pyramid, individuals with very high personal carbon footprints—the *super-polluters*.

Super-Polluters

Collectives of all stripes are the key to climate action. This has been the main argument of the book, along with highlighting how responsibility as members is usually the main responsibility of individuals when it comes to addressing climate change. Despite this, there are individuals who can bear direct responsibility for climate change, whom I call super-polluters. Super-polluters can potentially make a direct impact on the world's emissions with their own direct choices. More importantly, they wield so much (economic) power that they could choose differently if they wanted to. With some qualifications (which will come soon), super-polluters bear direct responsibility for their own emissions. In some cases, for super-polluters, reducing their individual carbon footprints is their primary climate change responsibility.

Many have argued against unilateral emission reduction responsibilities. For example, Baylor L. Johnson (2003, 272) asks: can someone who understands the problem of global warming and the contribution that cars make to it drive a large gas guzzler in good conscience? He argues, using the Tragedy of the Commons as his platform, that it would be a mistake to see as our primary obligation to reduce our individual burden on the environment unilaterally. Instead, we should seek a cooperation scheme to address the problem. I agree that it would not make sense to think that the primary responsibility of people is the unilateral reduction of their individual carbon footprint. As I have argued in the earlier chapters, we should push for collective solutions not only as citizens, but also as members of collective agents and constituents of unorganised collectives. However, I want to challenge the thinking that just because our responsibilities as members of collective agents or as citizens take priority, individuals would not also be responsible for re-thinking their consumption. These things do not cancel each other out: while promoting collective action and solutions is the most important thing we can do to combat climate change, our consumption norms and habits should also be critically evaluated, as I argued in the previous chapter. This will hold true especially for those individuals with very heavy carbon footprints.

With so much overconsumption and global inequality in emissions, it will always be somewhat arbitrary where to draw the line over who is a super-polluter and who is merely an above-average polluter, especially as the latter will also have carbon footprints several times over the carrying capacity of the Earth. There could be many ways to delineate the group and make the argument. I want to draw the line at a level where it is uncontroversial to claim that most of the individuals picked out by the category cause pollution at harmful levels. They also have so much economic room to manoeuvre that they have the freedom to choose differently. This does not mean that these individuals are free from societal pressures and social norms within their reference group; it only means that they are wealthy enough that they have genuine direct choices that can lead to large emissions or emission reductions. I suggest that super-polluters can be found among high-net-worth individuals (HNWIs), that is, people with at least US$1 million in investable wealth, excluding their primary residence and private assets.[4] It is estimated that there are currently 21.7 million people in the world that have this much investable wealth, according to the latest available data from 2022 (Capgemini 2023). This equals, very roughly, the top 0.3% of global emitters with the highest carbon footprints.[5] Super-polluters thus occupy the very tip of the emissions pyramid.

[4] Capgemini Financial Services, who compile the World Wealth Report, excludes primary residence, collectibles, consumables, and consumer durables from the investable assets. See www.worldwealthreport.com.

[5] I base this percentage on the figure that the top 0.1% in 2019 translates to 7,700,000 individuals (Chancel 2022). Still, the 0.3% is meant to be a rough estimate only, as there are different ways to calculate the global number of HNWIs.

For my purposes, it is safe to assume that while some people within this group might live frugally despite their wealth, and hence might even have average carbon emissions, most people within the HNWI group have lifestyles that lead to emissions on a scale that qualifies them as super-polluters. Some might of course already work to reduce their carbon footprints, and others might use their wealth mainly for philanthropic pursuits. Such things should matter in responsibility deliberations. We could even debate if all emissions from mega-pollution should be treated alike, or if the use of private jets is more acceptable when they are utilised for important political purposes, or to bring great joy to millions of music fans around the world as part of a stadium tour (as opposed to going on a private holiday in as extravagant manner as possible, for example). I set such issues aside here. As elsewhere in the book, I wish to underline that the actual responsibilities of any given individual are contextual. The point I want to make here is only that in our world in which carbon inequality is rampant, the pollution from the typical person occupying the top wealth tiers is clearly harmful and unsustainable.

HNWIs can be separated into three wealth bands: 'Millionaires Next Door' (US$1 million to US$5 million in investable wealth), 'Mid-Tier Millionaires' (US$5 million to US$30 million), and 'Ultra-HNWIs' (US$30 million or more), which also includes billionaires (Gapgemini 2023). Women account for only 11% of Ultra-HNWIs, although the proportion is gradually rising (Imberg et al. 2023). Note that the definition of HNWI used here is a more exclusive category than a millionaire, as only investable assets are included and not, for example, the primary residence, where most people's wealth is. Thus, the estimated 21.7 million HNWIs is a much smaller number than the estimate of millionaires that includes all their assets, the number of which was 62.5 million at the end of 2021 (Shorrocks et al. 2022).[6] The cities with the most HNWIs in 2022 were (in order of the highest number) New York City, Tokyo, The Bay Area (San Francisco and Silicon Valley), London, Singapore, Los Angeles, Hong Kong, Beijing, Shanghai, and Sydney (Henley and Partners 2023).[7]

[6] Credit Suisse also refers to them as HNW individuals, but note in their concepts and methods section that the definition of net worth used in their report encompasses 'the value of financial assets plus real assets (principally housing) owned by households, minus their debts. This corresponds to the balance sheet that a household might draw up, listing the items that are owned and their net value if sold' (Shorrocks et al. 2022, 19).

[7] Henley & Partners state that they refer to HNWIs as individuals with investable wealth of US$1 million or more.

Still, HNWIs can be found across the globe. For example, Dubai has 68,400 HNWIs, while Mumbai is home to 59,400 HNWIs, and Delhi boasts 30,200 (ibid.). Johannesburg has 14,600 HNWIs, while Cairo, Cape Town, Lagos, and Nairobi top off the top five HNWI cities in Africa (ibid.). However, the extreme wealth at the very top of the pyramid is still concentrated in countries that we are used to seeing ruling the global economy. Countries with the most Ultra-HNWIs are, in order of rank: the United States, China, Germany, Japan, the United Kingdom, Canada, Hong Kong, France, Italy, and India (Imberg et al. 2023).

Super-polluters have very high personal carbon footprints, meaning that the impacts from their direct consumption choices unquestionably cause harm. There has been a lot of debate about the moral significance of an individual's emissions and if it makes sense to discuss harm that individuals cause (see Chap. 1). These debates are based on the emissions of an average person in the United States or another wealthy country and concur that, while it is impossible to pinpoint the actual harm that individual emissions might cause, they will almost certainly cause harm and contribute to the risk of serious harms together with such emissions at various micro-thresholds (e.g., Broome 2016; Cripps 2016; Hiller 2011; Lawford-Smith 2016; Sandler 2011). It is important to note that the emissions figures from super-polluters are very large—so large that they do not fit the mould of such debates. While I have not been able to find estimates for HNWI per capita emissions, the global top 1% of emitters had personal carbon footprints of over 50 tonnes of CO_2 in 2021, against the global average energy-related carbon footprint of approximately 4.7 tonnes of CO_2 per person (Cozzi et al. 2023), while the per capita emissions of the top 1% in the United States have been estimated to be nearly 320 tonnes carbon equivalent (Chancel and Piketty 2015, 29). Recall that I am suggesting super-polluters are found within the top 0.3% of emitters, so their emissions will be significantly higher than these already very high figures for the top 1%.

If we accept individual direct responsibility concerning climate change harms, what could it mean in practice for individual super-polluters? I have suggested in earlier work (Hormio 2017) that we have a moral responsibility not to increase the probable risk of serious harm—the deprivation of fundamental capabilities—to other people, at least as long as we can do so at a less than significant cost to ourselves. This might demand quite a lot from super-polluters. After all, due to the wealth, they will have choices for low-carbon alternatives at no significant cost to themselves, including utilising all the latest clean energy technology. Super-polluters are also able to make lifestyle choices that have large impacts in terms of emissions, such as how large a home to live in or what modes of transport to use.[8] Such choices have cumulative impacts on the overall emissions.

Especially among the Ultra-HNWIs, lifestyle choices can have significant differences even at the individual level, as the personal carbon footprint of a billionaire can be enormous depending on their choices. For example, Roman Abramovich was calculated to emit at least 33,859 metric tons of carbon in 2018 (Wilk and Barros 2021), much of this down to his lavish consumption choices of owning a superyacht and private jets. Such extravagant lifestyle choices send strong signals about the material standard of living that individuals should aspire to. This is especially strong with influential people such as Abramovich or celebrities—the material standard of living that many of them advocate on social media, for example, is both unrealistic and unhelpful, even by aspirational standards. Excessive consumption is often pushed as the key to satisfaction and happiness in consumerist societies, even though there are many other ways of being. The very wealthy with all their options and choices could be in prime position to promote fresher, more interesting, and sustainable narratives about what makes life good. Of course, all of us should question what kind of signalling we are doing ourselves and if it makes sense to romanticise or aim for excessive consumption in the first place.

Despite lifestyles that include regular use of private jets, there is very little research into the consumption patterns or ecological footprints of the HNWI (Kenner 2015). As I noted earlier, for my purposes, it is safe to assume that while some people within this group might live frugally despite their wealth, most people within this group have lifestyles that lead to emissions on a scale that qualifies them as super-polluters. The primary reason for the very high per capita carbon footprint is the investments of the very wealthy (Chancel 2022). This is not surprising, as the category of HNWIs is calculated based on their investable wealth. I will discuss investments soon. But before that, in the next section, I will look at the other reason for the high emissions: individual consumption choices.

[8] These are to be separated from what Marshall (2014, 157) brands as 'petty lifestyle changes', such as carrying one's own eco-bags to avoid having to buy plastic bags.

Mega-Polluting Consumption

High disposable income tends to drive consumption, leading to higher emissions. As the household income increases, the proportion of expenditure in food decreases and the proportion of money spent on non-food items increases (Sri and Banerjee 2023). For the super-rich, their main sources of emissions are not things like the meat they consume (FAO 2013) but, rather, the various houses they own, how they travel between them, and more generally, their high-end travel choices, such as flying and yachting. Such choices manifold their footprints. Shue (2021, 141) labels such emissions as "morally intolerable", noting that momentary selfishness can cause millennial damage.

This will hold especially true for the top of the super-polluters, those among the Ultra-HNWIs who are billionaires. While the average person in the United States has the highest personal carbon footprint in the world, it is estimated that it would still take them 550 lifetimes to emit as much as the carbon footprint of the average billionaire (Wilk and Barros 2021). The Millionaires Next Door are of course unlikely to have mega-vast carbon footprints, but their lifestyles include many factors that bear very large carbon footprints. Such super-sized components of HNWI lives include owning very large homes, such as mansions, or several homes. Transport can include the use of helicopters, private jets, or other high-class travel. For example, chartering a private Boeing 747 emits 6500 tons of CO_2 per year (Wilk and Barros 2021). Such lavish lifestyle consumption choices lock super-polluters' emissions into a certain path.

The most conspicuous consumption comes from yacht ownership. Yachts are a status symbol—a clear sign to signal one's wealth and a way to convene one's economic and social status to others. A big part of their appeal is that most people could not afford them, even if they wanted to. Still, aside from mere pomp and show, there is a lot of human inventiveness at play in designing new fancy features for superyachts. Features that sound like fun include semi-submerged lounges with underwater observatories (nicknamed 'Nemo lounges'), outdoor cinemas created by projecting films on the lower sail, or slides that take you straight from the yacht to the sea (Zaltzman 2023). The same can be said for space tourism. Many people would love to see the Earth from space and experience the thrill of being launched into space in a rocket. There is no question as to why such activities are appealing, despite their very high carbon footprints. But those footprints are truly staggering. A typical superyacht (with a permanent crew, helicopter pad, submarines, and pools) emits approximately 7000 tons of CO_2 per year, which is against the average global carbon footprint of around 5 tons per person (Wilk and Barros 2021).[9] Space rockets emit 50–100 times the carbon per passenger than airplanes do, even for the very short flights they provide (Marais 2021). Because the emissions from an 11-minute space flight are at least 75 tonnes, they exceed the lifetime emissions of someone belonging in the poorest billion people on Earth (Weston 2021).

As the emissions from such high-end consumption choices are so large, it has been suggested that governments should try to reign it in through measures like progressive carbon taxes, targeted especially for excessive emissions (Chancel and Piketty 2015; Garcia and Stronge 2022). Choice editing is another potential route. The NGO sector has called for governments to raise taxes and/or to ban carbon-intensive luxury consumption, including mega yachts, private jets, and space tourism, arguing that they "represent a morally unjustified depletion of the world's scarce remaining carbon budget" (Gore 2021, 9). Some millionaires and billionaires have even publicly asked to be taxed more to help combat extreme inequality (Neate 2023). Such actions, like gathering support among your HNWI peer group to write an open letter and calling attention to it at a high-profile event like the World Economic Forum in Davos (as the millionaires and billionaires did), are examples of recognising a shared social orientation. It thus falls under shared responsibility discussed in Chap. 4.

While I have drawn attention to the direct responsibility of super-polluters, I do not suggest that they would not have other kinds of responsibilities as well, such as shared responsibilities as members and constituents of collectives. After all, even if your lifestyle emissions are excessive, it does not automatically mean that your direct responsibility will be your primary one. Many among the super-polluters will be among the most powerful individuals within their countries or internationally, and their shared responsibility as members of collective agents can be their most important responsibility. Think of top US lawmakers who are also HNWIs: their primary responsibility is dictated by their political role. In a similar way, many super-polluters hold top executive positions in collective agents, such as large corporations or finance entities, and their membership in these is their primary source of responsibility. This does not erase their

[9] Many owners rent their yachts when they are not using them, so renters can share some of the carbon footprint.

direct responsibility for their excessive emissions; it just means that their direct responsibility is not their primary climate-related responsibility.

It should be noted, though, that not all super-polluters occupy powerful roles within collective agents. Think, for example, of the family members (e.g., partners and children) of HNWIs. They might not occupy any powerful roles themselves within any collective agents, or even have that much money on their own, meaning that they might not even qualify as HNWIs. However, partners and other family members are customarily involved in high-end purchases, such as luxury cars, yachts, and private jets, regardless of the primary ownership of such assets (Imberg et al. 2023). If they consume at very high levels, their direct responsibility as super-polluters is their primary responsibility regarding climate change. If such super-polluters are underage, it would make sense to tally their consumption to that of their parents. Once they are adults and make decisions on their own, their super-polluting becomes their responsibility. If the super-polluters are also active on social media, they could potentially wield power as social media influencers who promote carbon-intensive lifestyles. For example, social media profiles and images of super-rich teenagers and young adults that are constructed around lavish consumption have the potential to contribute to social normalisation of emissions-intensive transport, such as high-end SUVs and air travel (Cohen et al. 2021). While greener consumption choices might influence the social norms around consumption within any social circles one occupies (see the previous chapter), such super-polluters with high-visibility social media accounts could have additional responsibility to contribute towards creating more sustainable consumption norms.

One might think that an easy way out for the super-polluter would be to offset their emissions. After all, many brands already offer products that they promise to be 'carbon neutral', in which the emissions of the product are offset. The general idea in offsetting is that we can essentially 'cancel' our own GHG emissions by paying for projects that reduce emissions somewhere else, such as building renewable energy infrastructure. Popular carbon offsetting involves measures such as reducing deforestation by paying countries to leave their forest untouched, as trees absorb carbon, or financing projects that diminish emissions elsewhere in the world (e.g., create sources of renewable energy or install efficient cooking stoves to replace cooking with firewood). When we offset, the theory is that the units of GHGs that we add to the atmosphere are met with an equal number of units that are subtracted from it.

It is difficult to prove that any investments in offsetting really deliver on emission-reduction promises, that is, that the reductions achieved by any given project are *additional* to what would have happened in the absence of the project. To qualify as a genuine carbon offset, the project cannot be one that would have been undertaken in the business-as-usual scenario. In other words, the investment secured through carbon markets must have been the deciding factor in making the project viable. If carbon credits would be awarded to projects that would have materialised in any case, they do not represent any genuine emission cuts. These could be projects that would have secured government funding or funding from private investors, even without the offsetting companies being involved. To give an example, ClimateCare—an award-winning reputable offsetting company—distributed 10,000 energy-efficient lightbulbs in a South African township as offsets. However, they soon discovered that an energy company was distributing the same kind of lightbulbs free to customers, including the township that ClimateCare had distributed their bulbs to, so the resulting reduction in emissions would have happened anyway (Davies 2007).

While additionality is a key concept in offsetting, it is hard to assess counterfactual claims about what would have happened without some project getting funding through carbon markets. This is especially so when you factor in any indirect effects that any projects might have, especially large-scale ones such as those of energy infrastructure. Indirect side effects are hard to measure but even harder to assess through counterfactuals. Even the direct effects prove a challenge, as it is inherently difficult to estimate the impact of any intervention from observational data. This is because it relies on estimating what would have happened in the absence of the intervention, which is especially challenging if the project takes place in a site where there have been other ongoing long-term conservation efforts (Guizar-Coutiño et al. 2022).

Apart from counterfactuals, there is uncertainty about what offset projects actually cause in the real world, including indirectly promoting harmful consuming practices (Hormio 2017). Offsets are calculated on projected outcomes, but projects may fail, no matter how well-intended and planned they are. In these cases, the reductions promised do not materialise, and offsetting has again failed to take place. Results can easily be overestimated. For example, one study found that carbon offset projects in the Brazilian Amazon consistently assumed higher deforestation in their crediting baselines than forest loss in synthetic control sites (West et al. 2020). Furthermore, reductions might be temporary. For example, trees planted for carbon offsets can end up burning in forest fires (Hodgson and Nauman 2021). Perhaps most disconcertingly of all, offsetting can create the impression that it is all we need to solve the emissions problem—and sends the misleading signal that no lifestyle changes are needed as buying cheap offsets is enough (Spiekermann 2014).

Investment Choices

Super-polluters have the potential to influence things not just through their consumption (or through the powerful membership roles they occupy) but also through their investments. In the inequality section at the beginning of this chapter, I noted that over 70% of the emissions generated by the global top 1% comes from their investments rather than their consumption (Chancel 2022). If super-polluters moved their personal investments away from fossil fuels and into things like building low-carbon buildings or developing new green energy technologies, they could make a significant impact on the global emissions outlook. Such investments could potentially bring about very good outcomes in terms of reduced emissions in other groups as well, as such solutions can be utilised widely once they can overcome the initial hurdle of getting funded.

At the moment, this is not the status quo. Recall from Chap. 2 how financial institutions, such as banks and hedge funds, are still enabling expansion of fossil fuel production by investing in the corporations or by lending them money for things like fracking or building new pipelines. While some banks have started to adopt policies against such destructive practices, they are in the minority. All in all, the financial sector and their equity holding portfolios are still deeply embedded in GHG-intensive assets (UNEP 2022, 67). The same can be said for super-polluters. Research by Oxfam (2015) found that between the Copenhagen (2009) and Paris (2015) climate conferences, "the number of billionaires on the Forbes list with interests in fossil fuel activities has risen from 54 in 2010 to 88 in 2015, while the size of their combined personal fortunes has expanded by around 50% from over $200bn to more than $300bn." There are thus very powerful personal interests at stake in fossil fuels. Unsurprisingly, the fossil fuel corporations are not the only ones engaged in lobbying for more time for their products, despite the accumulating harms they cause. Billionaires and their representatives are also active lobbyists, seeking to influence public policy and elections in their favour

(Hardoon 2017). Economic elites benefit from feedback loops that are formed when they increasingly control not only assets but also important organisations and networks. Financial elites can then use these positions to influence politics and policy processes, which leads to rules shaped in their interest, that is, which will benefit them financially (Green and Healy 2022). Having financial stakes in fossil fuels can mean supporting things like fossil fuel subsidies and blocking energy transition towards low-carbon alternatives by using their political influence (Kenner 2019). Some HNWIs will also have stocks in fossil fuels due to their high-level position as members of fossil fuel corporations, such as directors, executive vice presidents, and board members. The non-salary component of CEOs at large fossil fuel corporations is more valuable than the salary component (Kenner and Heede 2021). When your personal wealth depends on fossil fuels staying profitable, it gives you perverse incentives for increasing oil and gas production.

There are promising signals for global energy transition investment, however. Investments into the low-carbon energy sector were estimated at US$1.1 trillion in 2022, which includes renewable energy (wind, solar, biofuels, biomass and waste, hydropower), energy storage, carbon capture and storage, electric vehicles and charging infrastructure, heat pumps, hydrogen production, nuclear power, and sustainable materials (BloombergNEF 2023). This means that, for the first time ever, it was on par with investments into fossil fuels. Much more is needed though, as the speed of reversal of investments has been too slow. The investments need to triple to reach global net zero targets. Investments into energy transition and the grid needs to average US$4.55 trillion between 2023 and 2030 (ibid.).

It is in the long-term interest of the investors and the financial system itself to reduce investments in fossil fuels because a considerable share of these assets is likely to become stranded (UNEP 2022, 67). It has been suggested that the main investment objective of many HNWIs is to preserve their wealth for their descendants (Gapgemini 2023). It should thus be in the self-interest of super-polluters to invest in a more sustainable way, even when fossil fuels bring in short-term gains. China is the clear leader of the pack in global energy transition investment, accounting for nearly half of the investments, driven by renewable energy, electric vehicles, and steel recycling, with the EU in second place, if treated as a single block (BloombergNEF 2023). The new climate legislation in the United

States is expected to lead to rapid acceleration into these investments in the coming years (ibid.).

According to wealth management professionals working with HNWIs, concerns over climate change and other environmental and social problems have led to many investors seeking to increase their environmental, social, and governance (ESG) investments (Gapgemini 2023). The industry still needs clearer guidelines and transparency around these investments. I mentioned in Chap. 2 that there are legal cases against misleading ESG claims. There is also conservative backlash against ESG investments, including lawsuits in the United States (Setzer and Higham 2023).

It is important to include investments into the personal carbon footprints of the super-polluters, as otherwise we are missing the financial power that the HNWIs wield in addition to their consumption. Still, their decisions and responsibility as investors are not always as direct as their decisions and responsibility as high-end consumers. This is why their responsibility as investors might be better conceptualised as shared responsibility as members such as shareholders (Chap. 3), or as responsibility as constituents of an unorganised collective with a shared social orientation (Chap. 4) regarding making sustainable investments.

Although I have been focusing on the exorbitant influence and carbon footprints of a powerful few on the global emissions in this chapter, the discussion inevitably turns towards shared responsibility again. While calculating the personal share of company emissions based on shareholding can spark further discussion on the personal responsibility of board members in decarbonising their companies, a limitation of the approach is that the shareholding percentage does not necessarily translate to the degree of responsibility (Kenner and Heede 2021). The board members make decisions together, so issues related to responsibility as members apply here too, as discussed in Chap. 3. This intermingling of responsibilities is not surprising. As I have argued throughout the book, it would be artificial to discuss individual responsibility in isolation from collective and shared responsibility when it comes to climate change and other such large-scale issues.

Concluding Remarks

Organised, agential collectives are in a much better position to take climate action than individuals. Shared responsibility, how we act as members of collective agents, is usually our most significant responsibility regarding taking action on climate change. However, we should also discuss the direct individual responsibility of super-polluters. Their emissions are on a scale that is entirely unsustainable and goes against any urgent mitigation calls. Importantly, these emissions are unavoidable and often very frivolous. Due to vast carbon inequalities that exist in our world both between and within countries, there is a new urgency for ethics of individual emissions. In practice, such responsibility often intermingles with shared and collective responsibility.

References

Baer, Paul, Tom Athanasiou, Sivan Kartha, and Eric Kemp-Benedict. 2010. Greenhouse Development Rights: A Framework for Climate Protection That Is "More Fair" Than Equal Per Capita Emissions Rights. In *Climate Ethics: Essential Readings*, ed. Stephen M. Gardiner, Simon Caney, Dale Jamieson, and Henry Shue, 215–230. Oxford: Oxford University Press.

BloombergNEF. 2023. *Energy Transition Investment Trends 2023: Tracking Global Investment in the Low-Carbon Energy Transition*. Bloomberg Finance.

Broome, John. 2016. A Reply to My Critics. *Midwest Studies In Philosophy* 40: 158–171.

Caney, Simon. 2009. Justice and the distribution of greenhouse gas emissions. *Journal of Global Ethics* 5 (2): 125–146.

Chancel, Lucas. 2022. Global Carbon Inequality over 1990–2019. *Nature Sustainability* 5: 931–938.

Chancel, Lucas, and Thomas Piketty. 2015. *Carbon and Inequality: From Kyoto to Paris - Trends in the Global Inequality of Carbon Emissions (1998-2013) & Prospects for an Equitable Adaptation Fund*. PSE, November 2015.

Chancel, Lucas, Philip Bothe, and Tancrède Voituriez. 2023. *Climate Inequality Report 2023*. World Inequality Lab Study 2023/1.

Cohen, Scott, Hongbo Liu, Paul Hanna, Debbie Hopkins, James Higham, and Stefan Gössling. 2021. The Rich Kids of Instagram: Luxury Travel, Transport Modes, and Desire. *Journal of Travel Research* 61 (7): 1479–1494.

Cozzi, Laura, Olivia Chen, and Hyeji Kim. 2023. *The World's Top 1% of Emitters Produce over 1000 Times More CO2 Than the Bottom 1%*. Paris: IEA. www.iea.org/commentaries/the-world-s-top-1-of-emitters-produce-over-1000-times-more-co2-than-the-bottom-1.

Cripps, Elizabeth. 2016. On Climate Matters: Offsetting, Population, and Justice. *Midwest Studies in Philosophy* 40: 114–128.

Davies, Nick. 2007. The Inconvenient Truth About the Carbon Offset Industry. *The Guardian*, June 16.

FAO. 2013. *Tackling Climate Change Through Livestock: A Global Assessment of Emissions and Mitigation Opportunities.* Rome: Food and Agriculture Organization of the United Nations. www.fao.org/3/a-i3437e.

Gapgemini. 2023. *World Wealth Report 2023: Unlock Growth in Wealth Management.* Gapgemini Research Institute.

Garcia, Luiz, and Will Stronge. 2022. *A Climate Fund for Climate Action.* Autonomy Research Ltd. https://autonomy.work/portfolio/climate-fund-climate-action.

Gore, Tim. 2021. *Carbon Inequality in 2030: Per Capita Consumption Emissions and the 1.5^{0}C Goal.* Joint agency briefing note 5 November 2011, Oxfam International and the Institute for European Environmental Policy.

Green, Fergus, and Noel Healy. 2022. How Inequality Fuels Climate Change: The Climate Case for a Green New Deal. *One Earth* 5 (6): 635–649.

Guizar-Coutiño, Alejandro, Julia P.G. Jones, Andrew Balmford, Rachel Carmenta, and David A. Coomes. 2022. A Global Evaluation of the Effectiveness of Voluntary REDD+ Projects at Reducing Deforestation and Degradation in the Moist Tropics. *Conservation Biology* 36 (6): e13970.

Hardoon, Deborah. 2017. *An Economy for the 99%: It's Time to Build a Human Economy That Benefits Everyone, Not Just the Privileged Few.* Oxfam Briefing Paper, 16 January 2017. Oxfam International. https://doi.org/10.21201/2017.8616.

Henley & Partners. 2023. *The World's Wealthiest Cities Report 2023.* www.henley-global.com/publications/wealthiest-cities.

Hiller, Avram. 2011. Climate Change and Individual Responsibility. *The Monist* 94 (3): 349–368.

Hodgson, Camilla, and Billy Nauman. 2021. Carbon Offsets: A Licence to Pollute or a Path to Net Zero Emissions? *Financial Times*, August 31.

Hormio, Säde. 2017. *Marginal Participation, Complicity, and Agnotology: What Climate Change Can Teach Us About Individual and Collective Responsibility.* Helsinki: University of Helsinki.

Imberg, Maya, Maeen Shaban, and Nikoletta Szabo. 2023. *World Ultra Wealth Report 2023.* Altrata, Wealth-X. https://altrata.com/reports/world-ultra-wealth-report-2023.

Johnson, Baylor L. 2003. Ethical Obligations in a Tragedy of the Commons. *Environmental Values* 12: 271–287.

Kartha, Sivan, Eric Kemp-Benedict, Emily Ghosh, Anisha Nazareth, and Tim Gore. 2020. *The Carbon Inequality Era: An Assessment of the Global Distribution of Consumption Emissions Among Individuals from 1990 to 2015 and Beyond.* Oxfam International and Stockholm Environment Institute. https://doi.org/10.21201/2020.6492.

Kenner, Dario. 2015. *Inequality of Overconsumption: The Ecological Footprint of the Richest.* Global Sustainability Institute Working Paper No. 2015/2, November 2015. Cambridge: Anglia Ruskin University.

Kenner, Dario. 2019. *Carbon Inequality The Role of the Richest in Climate Change.* Abingdon and New York: Routledge.

Kenner, Dario, and Richard Heede. 2021. White Knights, or Horsemen of the Apocalypse? Prospects for Big Oil to Align Emissions with a 1.5 °C Pathway. *Energy Research & Social Science* 79: 102049.

Lawford-Smith, Holly. 2016. Difference-Making and Individuals' Climate-Related Obligations. In *Climate Justice in a Non-Ideal World*, ed. Clare Heyward and Dominic Roser, 64–82. Oxford: Oxford University Press.

Marais, Eloise. 2021. Space Tourism: Rockets Emit 100 Times More CO_2 per Passenger Than Flights – Imagine a Whole Industry. *The Conversation*, July 19.

Marshall, George. 2014. *Don't Even Think About It: Why Our Brains Are Wired to Ignore Climate Change.* New York: Bloomsbury.

Neate, Rupert. 2023. 'Tax Us Now': Ultra-Rich Call on Governments to Introduce Wealth Taxes. *The Guardian*, January 18.

Newell, Peter, and Mohamed Adow. 2022. Cutting the Supply of Climate Injustice. *IDS Bulletin* 53 (4) 'Reframing Climate and Environmental Justice': 31–46.

Newell, Peter, Freddie Daley, and Michelle Twena. 2021. *Changing Our Ways? Behaviour Change and the Climate Crisis: The Report of the Cambridge Sustainability Commission on Scaling Behaviour Change.* Cambridge Sustainability Commission.

Oxfam. 2015. *Extreme Carbon Inequality: Why the Paris Climate Deal Must Put the Poorest, Lowest Emitting and Most Vulnerable People First.* Oxfam Media Briefing, December 2.

Sandler, Ronald. 2011. Beware of Averages: A Response to John Nolt's 'How Harmful Are the Average American's Greenhouse Gas Emissions?'. *Ethics, Policy & Environment* 14 (1): 31–33.

Sengupta, Somini. 2023. The American exception. *The New York Times Climate Forward Newsletter*, February 28.

Setzer, Joana, and Catherine Higham. 2023. *Global Trends in Climate Change Litigation: 2023 Snapshot.* London: Grantham Research Institute on Climate Change and the Environment and Centre for Climate Change Economics and Policy, London School of Economics and Political Science.

Shorrocks, Anthony, James Davies, and Rodrigo Lluberas. 2022. *Global Wealth Report 2022: Leading Perspectives to Navigate the Future.* Credit Suisse Research Institute.

Shue, Henry. 2021. *The Pivotal Generation: Why We Have a Moral Responsibility to Slow Climate Change Right Now.* Princeton: Princeton University Press.

Spiekermann, Kai. 2014. Buying Low, Flying High: Carbon Offsets and Partial Compliance. *Political Studies* 62: 913–929.

Sri, Priyanka, and Rangan Banerjee. 2023. Characteristics, Temporal Trends, and Driving Factors of Household Carbon Inequality in India. *Sustainable Production and Consumption* 35: 668–683.

UNEP. 2022. *Emissions Gap Report 2022: The Closing Window — Climate Crisis Calls for Rapid Transformation of Societies.* Nairobi: United Nations Environment Programme. www.unep.org/emissions-gap-report-2022.

West, Thales A.P., Jan Börner, Erin O. Sills, and Andreas Kontoleon. 2020. Overstated Carbon Emission Reductions from Voluntary REDD+ Projects in the Brazilian Amazon. *PNAS* 117 (39): 24188–24194.

Weston, Phoebe. 2021. 'Luxury carbon consumption' of top 1% threatens 1.5C global heating limit. *The Guardian*, November 5.

Wilk, Richard, and Beatriz Barros. 2021. Private Planes, Mansions and Superyachts: What Gives Billionaires Like Musk and Abramovich Such a Massive Carbon Footprint. *The Conversation*, February 16.

Zaltzman, Julia. 2023. Forget a Helipad. These 10 Bonkers Superyachts Have Everything from Snow Rooms to Golf Tees. *Robb Report*, June 29, robbreport.com/motors/marine/gallery/9-bonkers-features-superyachts-1234861620.

CHAPTER 6

Why We Need Ethical Arguments to Set Good Climate Policies

Introduction

As the book has argued, climate change is an urgent and severe threat that needs an immediate and wide response, so much so that many are already calling it a climate crisis. Urgent action is required on a large-scale, but mitigation is often presented by policymakers as an economic issue: what would strong regulation to curb emissions do to our economies, what are the most cost-efficient switches, and so on. The problem with this framing is that it looks like the whole issue is just an optimisation puzzle waiting to be solved, albeit a highly complex one. Yet many normative questions without easy answers arise. For example, some philosophers have argued that the rich nations responsible for most of the emissions should, as a matter of equity or corrective justice, bear the brunt of the mitigation and adaptation costs (Gardiner 2016; Shue 1999; Vanderheiden 2008), or that distributive justice means that poor nations should be allowed to emit more (Singer 2002). Some legal theorists have responded that these kinds of normative arguments are utopian and suffer from problems of feasibility (Posner and Weisbach 2010). In their stead, they advocate treaties that take national self interest seriously (Weisbach 2016). According to such views, ethical arguments do not have a role to play in setting climate policies.

This chapter argues that there is a vital role for normative philosophical arguments in climate policy and economics. Even though ethical theories

S. Hormio, *Taking Responsibility for Climate Change*,
https://doi.org/10.1007/978-3-031-51753-2_6

cannot offer comprehensive policy proposals and straightforwardly tell us what to do in complex situations, they nonetheless have an important role in climate policymaking. Namely, they can highlight what values are at stake and help ground normative arguments in public deliberations. After all, ethical questions are embedded in policy choices and these value choices should be debated openly. Importantly, ethical arguments can also inspire new horizons of thought.

The function of ethical arguments in climate policy is also relevant to the wider question of what decision-making should look like at a time of climate crisis. The slow progress so far at the United Nations level and in many national climate policies alike has made many sceptical of the power of democracy to deal with climate change in the first place (e.g., Jamieson 2021; Shearman and Smith 2007). This is partly due to the intergenerational aspects and the long timeframe of seeing benefits from mitigation policy efforts, resulting in lack of incentives for politicians operating in short-term election cycles (Schmidt 2021). Some commentators worry that climate policy led by bureaucrats and state elites has made the issue seem remote and alienated engagement at the local level (Stevenson 2013). Others argue that we are past the point of creating better institutions and policy due to decades of inaction and delays in climate change mitigation, with some even advocating eco-authoritarianism (Beeson 2010; Hickman 2010). To avoid such an outcome, it has been argued that we might need not just increased localisation of democracy, but also large-scale transition and relocation into ecovillages, eco-neighbourhoods, and other such eco-communes (Fischer 2017). Be that as it may, these challenges bring us back to the importance of structures, which has been a running theme throughout the book. The lack of meaningful action can be seen as a failure of an over-reliance on technocratic solutions—along with populist politics (Huber 2020) and the distorting impact of the powerful fossil fuel lobby (Oreskes and Conway 2010)—rather than a failure of democracy per se. This is not to deny that democratic systems should be reformed and strengthened to be better able to address the climate challenge (Willis et al. 2022). When citizens get a more direct say, this can result in better policies, like the Ecuadorian direct democracy example (mentioned in Chap. 2) of banning oil drilling in the Amazon shows, or the example of the Irish Citizens' Assembly discussed later in this chapter. This is not inevitable, as nothing is in social matters, but it is entirely possible.

Contrary to those who think that due to the urgency and scale of the threat we need more purely technical solutions or even authoritarian regimes, I contend that more genuine deliberation about values is required to address the crisis. Deliberative democracy can help to guide climate change policy by integrating different perspectives on an issue, creating collective values to solve common problems, and enable new ethical relations to emerge by expanding the participants' thinking (Blue 2017). On top of these benefits, deliberation can also help to organise the voices of the marginalised (Dryzek and Niemeyer 2019). It can even help to forge public commitment for action that takes the long-term consequences of our decisions and their impacts on future generations seriously (MacKenzie et al. 2023). However, although I discuss participation and the interaction between experts and citizens, my goal is not to develop a novel conception of how to make climate change policy more democratic or offer a new line of defence for participatory politics. Rather, I aim to explore the important role for ethical arguments in setting climate policy and discuss why, despite this, there is no bypassing the occasionally messy political debates.

I begin by looking at how ethical questions always come into economic models used in decision-making and argue that the idea of neutral technocratic policymaking on large-scale issues such as climate change is fiction. I understand technocracy to mean decision-making based on scientific and technical expertise instead of open political debate based on values. We need to be alert whenever normative positions masquerade as neutral solutions. Simply debating the values openly, but only between technocrats and other experts, is not a solution either, even though the old Platonian idea of epistocracy as a supposedly better option over democracy (Brennan 2016) has gained some traction lately. In addition to strong egalitarian objections to epistocracy, the claim about the better outcomes it would bring can be questioned as well, as experts are also prone to fall into traps of biased thinking and narrow framing of issues. I back up this claim by discussing how the framing chosen by technocrats has led climate policy astray so far and discuss the epistemic benefits of deliberative and participatory democracy.[1] I then argue that self-interest, viewed by some as an alternative to ethics in climate policy, is insufficient in itself for guiding policymaking.

[1] My argument therefore shares elements with those of epistemic democrats, in particular in respecting the judgement of ordinary citizens (Schwartzberg 2015).

I should make it clear that although I argue against primarily technocratic climate policy, I am not denying the importance of technocratic expertise, nor am I in any way against experts. On the contrary, both climate science and the Covid-19 pandemic have underscored the need to listen to experts in policymaking.[2] They also highlight the grave dangers of populist politics that want to shun or sideline science. But just as the pandemic made it clear how valuable and important contributions by epidemiologists and other experts are, it also highlighted the need for political decisions based on values. There is no formula for calculating the loss of lives against economic costs, if this is indeed the trade-off. Difficult political decisions need to be made. Due to the nature of climate change, we need many technocratic solutions, including treaties to ease international cooperation. However, I contend that they should not and cannot be the whole solution. Ethical questions should be debated openly in order to get climate policy right.

No Formula

I will focus on climate economics in this section. Normative considerations always come into the calculations that economists prepare for policymakers as background material for their decision-making. When it comes to major public policy decisions, the values embedded in economic models should be made transparent and they should be deliberated publicly.

Climate change has not been met with the urgency it requires, and decades have been lost through insufficient action. An influential policy paper that contributed to the political inertia around climate change in the United States after the problem became known was the 1983 report *Changing Climate: Report of the Carbon Dioxide Assessment Committee* by the National Research Council. It gave politicians the feeling that instead of prompt action and tough decisions, all they needed to do was to fund more research (Oreskes and Conway 2010, 174–183). It comprised five chapters written by natural scientists and two by economists. The chapters by natural scientists concluded that human activities had increased carbon

[2] Covid-19 also illustrates the role of debate among experts in rapidly changing situations in which decisions have to be made quickly under conditions of uncertainty. Despite the urgency of climate change, the decision-making timeframe looks different: we are not faced with the threat of exponential growth within days or weeks.

dioxide concentration and that this would affect ecosystems, the weather, and agriculture. Mitigation was offered as the solution. The economists, however, had a different message: nothing more than a wait and see approach needed to be taken. The synthesis of the report concentrated on the economists' message and largely ignored the natural scientists (National Research Council 1983).[3]

The economists advocated a wait and see approach based on discounting. The assumption in their models was that serious changes were distant enough to be discountable. Discounting is a standard tool of economic analysis that allows effects occurring at different future times to be compared. First, all aggregate costs and benefits are expressed in terms of their present value. Then the discounted values are compared to each other. A policy is considered desirable if its net present value is positive. The discount rates used in public policy can vary a lot but are typically in single figures. However, when it comes to long-term policy decisions, as is the case with climate, a change of a few percent can have enormous effects. This is because the timeframe is so long. To give an example (from Davidson 2017, 22), in two hundred years' time, a thousand dollars would be valued at $137 today if a discount rate of 1% were chosen, whereas at a discount rate of 5% it would be a measly $0.06. Compared to a thousand dollars, both sums are very low, illustrating how even a low discount rate over a long period of time always leads to future utility being valued in very meagre terms, effectively reducing distant benefits to zero.

These relatively small differences in the consumption discount rate make for big differences in terms of policy recommendations. To give a prominent example, the Stern Review on the Economics of Climate Change utilised a low discount rate of 1.4% and concluded that we should take immediate action to reduce emissions. If no action is taken, the report warned that the overall costs and risks of climate change will correspond to losing at least 5% of global GDP each year (Stern 2007). In contrast, William Nordhaus's (2007) analysis assumes a higher preferred consumption discount rate (4.3%), which results in much less aggressive mitigation

[3] Sidelining the concerns of the natural scientists in the synthesis report used by policymakers was not accidental, but part of a systemic effort to delay action on climate change, backed by the fossil fuel industry and ideological concerns (Oreskes and Conway 2010, 169–215).

action being required now.[4] Nordhaus (2007, 695) argues that Stern's approach is inefficient because "it invests too much in low-yield abatement strategies too early", whereas an efficient strategy would invest in conventional capital at the beginning and then use those resources to invest heavily in 'climate capital' later on.

Discounting future utility is a standard practice in economics, but it is problematic when applied to climate change (and other very long-term issues) due to the timescale involved and the uncertainties that this entails. Although economic analysis tends to lead policy discourses over climate change, Broome (1992, 19) has argued that cost-benefit analysis amounts to self-deception in this context:

> Since governments must act, research on intergenerational relations must be aimed ultimately at providing guidance on how to act. Nevertheless, I believe it would be wrong to adopt the narrow aim of developing some formula for cost-benefit analysis, which governments could simply apply. I shall not confine myself to deriving a discount rate from current economic theory. The uncertainties of the problem are enough to make that exercise pointless. Cost-benefit analysis, when faced with uncertainties as big as these, would simply be self-deception. And in any case, it could not be a successful exercise, because the issue of our responsibility to future generations is too poorly understood, and too little accommodated in the current economic theory.

Even if economic theory could accommodate the issue of our responsibility to future generations, discounting would still not become a value-free, neutral policy tool. Discounting makes several normative assumptions (Hormio 2017), such as that future generations will be considerably better off than us, and that natural capital (e.g., ecosystem services such as clean drinking water) is substitutable by other forms of capital (from education to various material goods). The growth optimism inherent in discounting is rendered particularly problematic by the possibility of a runaway climate change scenario. With all feedback mechanisms still not

[4] Goulder and Williams III (2012) argue that debates like this blur the differences between behavioural and social welfare functions in optimisation models. The former is about how individuals *would* behave under various conditions and the latter about how individuals or societies *should* behave. They propose that two rates should be separated that would distinguish the desirable from the actual: a social-welfare-equivalent discount rate (appropriate for determining whether a given policy would augment social welfare) and a finance-equivalent discount rate (suitable for determining whether the policy would offer a potential Pareto improvement).

properly understood and not all tipping points known to us, our climate might become unstable much sooner than previously thought.

I should make it clear that my argument does not only consist of observing that there is no simple formula for policymakers because discounting and cost-benefit analysis always assumes values and a normative perspective.[5] This point has already been made by many economists, who caution against seeking simple policy answers from economic analysis, even when there is pressure to provide such answers. For example, Martin Weitzman (2011, 289) has commented on the demand for policy recommendations that "Practical men and women of action have a low tolerance for vagueness and crave some kind of an answer, so they have little patience for even a whiff of fuzziness from two-handed economists" (this is in reference to a quip attributed to former US President, Harry S. Truman: "Give me a one-handed economist. All my economists say, 'on one hand ... on the other.'") Another example is provided by Nordhaus (2007, 701) who writes: "No sensible policymaker would base the globe's future on a single model, a single set of computer runs, or a single national or ethical perspective. Sensible decision making requires a robust set of alternative scenarios and sensitivity analyses." This kind of cautioning still leaves the door open for purely technocratic decision-making, as long as several models are considered and different experts listened to. In contrast, my argument in this chapter is that purely technocratic decision-making will not suffice when it comes to climate policy. This is both because it can lead to epistemically harmful things such as too narrow framings, as I will argue in the next section, but also because debate over values cannot be bypassed, which I will argue later in this chapter.

Still, I want to again underline that my argument is not that we should get rid of technocrats. They are important for setting good and efficient policies. The argument is only a rejection of purely technocratic climate decision-making in the light of the inescapability of values embedded in climate policy. Climate policy questions are so complex that many trade-offs need to be made and conflicting interests balanced, meaning that there are no simple ethical solutions to these challenges (Kowarsch and

[5] Even the most sophisticated cost-benefit analysis cannot incorporate all of the things we value. Methods like deriving estimates of willingness to pay for things outside the market face many problems as reliable indicators of our preferences. Things that we tend to value very highly, like the lives of loved ones, cannot be incorporated into a cost-benefit analysis as they cannot be assigned a determinate value. As Steven Kelman (1981) has argued, things designated infinitely valuable or 'priceless' lose their value-affirming and value-bestowing meanings if we try to replace the description of pricelessness with even a very high valuation.

Edenhofer 2016). The discussion on discounting and cost-benefit analysis in this section has offered just one example of why this is so. Instead of formulas, what is required is public deliberation, because we need to be honest about the normative nature of things like setting a discount rate. As Amartya Sen (2007, 29) cautions, it is "crucial not to reduce important issues of human evaluation, which demand reflection and deliberative social assessment, into narrowly technocratic matters of formulaic calculation". He reminds us that what is really at stake in the debates around discount rates is "social evaluation of gains and losses over time", and that fundamentally this is "a deeply reflective exercise and a matter for public deliberation, rather than one for some kind of a mechanical resolution on the basis of some simple formula". These ethical claims need to be weighed in political debate.

Narrow Framings

I argued in the previous section that we need to be alert whenever normative positions masquerade as neutral technocratic solutions. Narrow technocratic framing can also lead us to look for solutions in the wrong places, as I will argue next. That is why technocracy needs participatory democracy to balance it and keep it in check, as people can correct each other's errors and bring unique viewpoints to the table. I do not deny the intrinsic worth of the values that are embedded and embodied in democratic processes, such as equality, but here I focus only on the epistemic strengths.

When it comes to climate change, the problem has been framed unsuccessfully since the beginning. George Marshall (2014, 162–167) describes how decision-makers and policy strategists drew on what they saw as precedents—including the successful efforts to create a binding international treaty to deal with ozone depletion—and made assumptions based on perceived similarities which, on closer inspection, would not hold.[6] This "cognitive error on a vast scale" resulted in "ever-more-energetic confirmation bias" that resulted in the same mistakes being repeated over and over again, and "created an optimistic narrative of resolution and renewal

[6]The challenge with ozone depletion was relatively simple: to develop substitutes for chlorofluorocarbons (CFCs) used in aerosols and refrigeration. The chlorine included in CFCs destroys ozone, allowing harmful ultraviolet radiation to come through. The effects were felt more locally and the solution was not very costly. However, one of the replacements, hydrofluorocarbons, are very potent greenhouse gases, but a deal to phase these out was agreed in 2016.

that was entirely inappropriate for the irreversible and open-ended problem of climate change" (2014, 166).

> Frames do not just focus the attention: they define the areas for disattention. These precedents bound climate change to a limited set of meanings that actively excluded other approaches. They defined climate change as an environmental issue and therefore not a resource, an energy, an economic, a health, or a social rights issue. They determined that it would be best managed through emissions trading, and therefore not through regulation, taxation, and rationing. And the U.N., glowing from the success of its process to prevent ozone depletion, determined that climate change would be best controlled through international protocol rather than regional or multilateral agreements. (Marshall 2014, 166–167)

Economist Thomas Schelling argues that the way to simplify the "awfully complicated hodgepodge" of our current climate change policies would be "to put the cap on the fossil fuels, not on different industries—a cap on oil and gas at the wellhead, a cap on coal at the Minehead" (interview in Marshall 2014, 170). In other words, rather than concentrating on regulating greenhouse gases and creating emissions trading systems, the focus could be on what produces these gases and emissions in the first place. This could mean taxing the carbon content in the fossil fuels directly or constraining the development of new sources of fossil fuels.[7] Yet this has never been debated at policy level, and the Intergovernmental Panel on Climate Change does not present data on fossil fuel production, only on greenhouse gas emissions. The discussion to concentrate efforts on tailpipe emissions, rather than at the wellhead, has not even been successfully lobbied by the fossil fuel industry: they did not generate the dominant approach (although they have since been involved in upholding it). Somewhat puzzlingly, within the circle of international climate negotiations, a debate to question the tailpipe approach never took place.[8] Rather, the framing seems to have just happened. Why? Marshall (2014, 173) suggests that we should understand it primarily "as an extreme error of

[7] This is not to deny the possible practical virtues within the current carbon pricing system in terms of transparency or integration in accounting approaches, for example, but such things are dependent on the initial political framing and the choice to not to focus on production.

[8] Marshall (2014,171) writes that "there were no fights, no struggles, no backroom deals. There did not need to be because it was never discussed". He bases this claim on interviews with people involved in climate negotiations.

judgment resulting from cognitive error and false categorization", namely that,

> Scientists categorized climate change as a tailpipe issue because production was considered a political issue that was outside of their domain. Policy makers then categorized climate change as a tailpipe problem because they drew on recent available experience that suggested viable solutions to tailpipe problems. Confirmation bias and a socially constructed norm of disattention finished off the job.

Hence, putting controls on the production of fossil fuels is not discussed in national or international politics because questions concerning the wellhead do not exist within the dominant debate. This allows most Western governments to have simultaneously enacted programmes to subsidise renewable energy production, while at the same time continuing to encourage and subsidise ever-larger investments in developing new sources of fossil fuels (Marshall 2014, 173). Needless to say, the latter has far exceeded the former. As the key players involved seem to be too embedded in their narrow institutional frameworks to see the dissonance of this approach, the thinking must be challenged from the outside.

I think that part of the solution to misleading framings that result from expert biases will be allowing more voices to be heard in decision-making within the global context of the climate crisis and placing more power in local contexts.[9] The idea is not to subject all climate-related policymaking to public deliberation. That could make decision-making incredibly slow and complex, not to mention ineffective and costly (although deliberation does not automatically entail such things). The idea is that those decisions that come with novel value choices should be subjected to participatory deliberation through methods like minipublics. With this I mean to refer to value choices that are novel due to the rise of either some new situation (like a new technology) or a new configuration of

[9] Public participation has been enshrined in the United Nations Framework Convention on Climate Change (UNFCCC) from the beginning. Despite the rhetoric, participatory and deliberative approaches have never taken a central position in the UNFCCC negotiations. We should not conflate being consulted and invited to provide input with having an actual say in the process (Schroeder 2010). However, even with unbalanced representation, it has been argued that the UNFCCC process helps different actors to come together, and the meetings have an important facilitative function in holding the polycentric regime together (Lövbrand et al. 2017).

circumstances, that has no clear precedents. Once an overall plan for how to deal with the issue has been agreed upon, the value choices that trickle from the overall plan can be agreed upon by experts alone, as long as they follow from the top-level choices. So, for example, once a region has decided to transition to a form of renewable energy or adopt carbon taxes, the details and technological fine-tuning of the plan can be done at the administrative expert level, as long as there are no novel value choices that arise during the process.

While debating the merits and challenges of participatory and deliberative approaches to climate change in detail falls outside the scope of this chapter (for that, see Blue 2017; Fischer 2017; Willis et al. 2022), I will discuss their benefits in general terms. Participatory approaches expand the available policy inputs, thereby contributing to more robust and informed decision-making (Blue 2017). When we reason together by engaging in an open debate, our biases and convenient rationalisations are easier to identify and to challenge. Empirical research suggests that as long as they engage in debate, groups reason better than individuals (Moshman and Geil 1998). The debate should be genuine in the sense that the group members actually hold divergent views as they begin their debate, because heterogenous groups are better at counteracting biases in order to determine the best alternative (Schulz-Hardt et al. 2002). Cognitive diversity allows for the emergence of collective intelligence (Landemore 2013).

None of this, by itself, is an argument against technocratic decision-making: put enough experts with heterogenous views together, and you might get to reap the benefits of debate-based group reasoning. Expertise-based group rationality is an important part of how democracy works efficiently, but it cannot be the whole story. This is in large part because—for many difficult questions—there is no optimal solution, only better and worse ones. While experts can weed out the bad solutions, we need democracy to work out which among the better options to choose from (see also Chap. 1). When choices are about values, it calls for open debate. After all, values are inherently plural and they depend on our personal histories. Our values are shaped by our relationships and the contexts we have been exposed to, and our preferences and values can change over time and in different contexts (Anderson 2003). As our positions invariably make us more sensitive to some pieces of information than others, in decision-making it is important to have as many perspectives represented as possible. Deliberative approaches will widen up the range of viewpoints and can help to foster new ideas and paths for climate policy.

Participatory and deliberative democracy is of course not a new invention: ancient Athens adapted more quickly to new challenges and a changing environment than its major competitors partly due to engaging in participatory processes (Ober 2008). The participation of non-elite citizens in government was a major public investment, but it paid off through encouraging policy experimentation and enhancing levels of social cooperation. By being able to utilise social, technical, and latent knowledge among their citizens, participatory and deliberative democracies have a competitive edge over authoritarian regimes relying heavily on like-minded experts (Ober 2008). Utilising widely dispersed knowledge demands resources, but it fosters innovation and learning. Through the mechanism of inclusive deliberation, decisions that have higher epistemic value can be created, even when the epistemic quality of the input of an individual citizen is limited (Landemore 2013).

Advocates of deliberative approaches also argue that when we enter into a deliberative process, this creates different forms of civic virtue. In a deliberative setting, the participants need to pull together in order to resolve shared problems, and the values that emerge from this can help us to prioritise collective interests over material self-interest (Blue 2017). In other words, democratic deliberation and discussion can shape and evolve our preferences. Through engagement in democratic processes, "preferences are not simply aggregated but informed and transformed in relation to each other" (Benson 2019: 78). Instead of purely abstract, theoretical, independent thinking, rationality born out of democratic deliberation is an embodied and situated social activity. Deliberation and debate should be understood broadly here to include not only centralised public spheres and forums, but also the various networks of governance and communications in our interdependent world, comprising of states, different organisations, individuals, and markets (Stevenson and Dryzek 2014). However, although the concept of new values that are created together is important, I do not suggest that all deliberations can or need to end up in such a harmony. Climate change is a big and multifaceted issue, so there is bound to be disagreements and alternative paths that come up in deliberations.[10] This is fine: there is value in democratic disagreement, and it can create brand new paths forward.

Some worry that participatory democracy will result in an even worse policy than the one we have achieved so far when it comes to climate change.[11] If people are allowed a say in matters of global and intergenerational scope, how are we to ensure that their own short-term interests and local biases won't cloud their judgements? For example, what mining community would choose to sacrifice their fossil fuel-dependent employment, even if this were to improve the overall national or global emissions picture? I think the answer to this concern is that the level for the debate has to be chosen carefully. The local level could be the right one for decisions such as participatory budgeting for the use of adaptation funds, or how the need for more mitigation measures should be confronted in that community, but not for making decisions on renewable energy infrastructure for the country, for example. This is not to say that it is easy to draw the line on what the right level is on a given issue. More generally, there exists a debate about widespread assumptions inherent in the idea of different levels of politics. Perhaps local and national level political debates should be more combined, and citizens could even be connected to global governance through innovations in deliberative democracy (Niemeyer 2020). In any case, wider policy decisions should employ direct democracy or participatory methods and the minipublic must draw on different groups of people and geographical locations, with the sessions facilitated

[10] As climate change affects communities in different ways, in many cases there will be many valid perspectives about the best way forward. Commentators have argued that we need a new framing based not only on climate science (and the supposedly objective course of action that derives from it), but one that also includes local people and places (Nightingale et al. 2020). This can include various strategies, such as public outreach outside academia, or coproduction of knowledge between academics and grassroots organisations and groups (Neimark et al. 2019). To give an example, combining Western climate science with ecological knowledge held by Indigenous people has highlighted the importance of different notions of time, and how policies based on Western calendar can make it harder for Indigenous communities to adapt their hunting and fishing practices to the changing seasons (Chisholm Hatfield et al. 2018). Such inclusive ways of going forward with climate policy could also be best placed to address existing inequalities that are exacerbated by climate change. These inequalities are on many dimensions, making the ethics of climate change more complex as the effects intensify (Mittler 2014).

[11] This is the most common objection I have gotten when I have presented on deliberative approaches to climate policy.

carefully. The danger in facilitation is that it guides the participants too much, down the path of some pre-chosen 'correct' decision (Lepori 2019), but transparency in the process and active engagement by civil society actors can alleviate such concerns.

In any case, there is growing evidence that deliberation within mini-publics can reduce partisan divisiveness and facilitate sophisticated reasoning that transfers values into policy preferences (Dryzek and Niemeyer 2019). One example is the Irish Citizens' Assembly, who met between October 2016 and April 2018 (Coleman et al. 2019). The Assembly, consisting of a chairperson and 99 citizens, met over 12 weekends to discuss 5 policy issues, including climate change, and what Ireland's climate change policy should look like. Members were chosen randomly, representing different ages, social classes, location, and gender. After being briefed by experts, the members debated the issues in groups, before voting on what to recommend to a special parliamentary committee, which was established to take forward the assembly's recommendations on climate policy. The parliamentary committee put together a report that was credited with significantly shaping Ireland's ambitious Climate Action Plan, published in June 2019.

As noted in the introduction to this chapter, my aim is not to deny the importance of expertise in decision-making, and I do not advocate that we should shut out technocrats from policymaking. That would lead to worse decisions and probably quite chaotic policymaking on the large-scale issues. Instead, I advocate that technocratic decision-making should be complemented by public participation and deliberation to make policies more robust and inclusive. Strengthening and enhancing democracy is required to meet the climate crisis (Willis et al. 2022).

Self-interest and Ethical Reflection

Not everyone is convinced that we need open ethical debate on climate policy. Some commentators consider that ethical arguments are highly likely to divide people even further, so we should try to focus climate policy on a narrower stance. To anticipate possible objections, I will discuss legal theorist David A. Weisbach (2016) as an example of the anti-ethics stance. He suggests self-interest as an alternative to justice and ethics in climate policy. I will first discuss self-interest in policymaking, before turning my attention to the role of ethics and normative philosophical arguments in policy debates. I will argue that because self-interest is not the whole story, but just one value among others, we need deliberate democracy to set policy.

In his debate with philosopher Stephen M. Gardiner in their jointly authored book *Debating Climate Ethics* (2016), Weisbach claims that a climate treaty that goes against the perceived national self-interest of the countries of the Global North has little or no chance of succeeding (Weisbach 2016, 144–145). He argues that to obtain a climate treaty that is effective, measures designed to redistribute wealth must be excluded, along with any measures that address historical wrongs against countries in the Global South.[12] Unmitigated climate change will end up hurting the poor the most, so obtaining a workable treaty should be a priority.[13] I will focus on the way Weisbach views pursuing self-interest as something that can replace ethical arguments and reflection. The way he approaches self-interest throughout the book gives one the impression that considerations of self-interest almost automatically point to some kind of an optimal solution. For example, he argues that,

> To a great extent, if we are behaving in a way that is contrary to our own self-interest, we do not use, or need, ethics to tell us what to do. We don't say, "Stop hitting yourself in the head with a hammer. It is unethical." We say, "Stop hitting yourself with a hammer. You will hurt yourself." The appeal is purely to self-interest. (Weisbach 2016: 152–153)

As I see it, there are three problems with what I will call *the pure self-interest approach*. The first is the importance of the context. Perhaps it is true that in certain contexts what is in our self-interest is so obvious that it requires no debate or ethical considerations. If this is so, it looks like hitting yourself in the head with a hammer qualifies: there is just no plausible context in which doing so does you long-term good despite the pain, so it

[12] Gardiner (2016, 109) claims that it remains an open question as to whether climate justice should be isolated from the wider concerns of global ethics. He finds the idea of a two-track approach far-fetched: why would powerful nations who resist climate justice endorse a robust global welfare or foreign aid treaty (Gardiner 2016, 94–95). Gardiner also notes how the focus on China, India, and Brazil obscures the plight of the most vulnerable countries.

[13] I agree with Weisbach that all big emitters, including China and India, should avoid locking themselves (or other countries) into the fossil-fuel energy infrastructure. Although China is leading the pack in investments in renewable energy, they are also heavily reliant on coal. Chinese companies are not only building coal plants on Chinese soil, but also in other countries, such as Pakistan, Vietnam, Egypt, and Malawi (Tabuchi 2017).

is clear that you are not acting in your self-interest. But how applicable or similar is this context to climate change policy? It seems that the answer is: Not at all.[14]

This brings us to the second problem with the pure self-interest approach: it jumps straight from the individual level to the collective level. In most cases, Weisbach employs self-interest to discuss what collective actors such as nation-states and governments should do. There he links self-interest to mainly economic concepts such as cost-effectiveness and cost saving, or avoiding significant job losses or trade disadvantages (Weisbach 2016, 158–160). But how could self-interest work in the 'stop using the hammer' kind of way when we discuss collectives? Government policy is about balancing the interests of different groups, and a policy proposal usually ends up advantaging and disadvantaging different individuals. To balance these interests, we need politics, in which ethical considerations also come into the mix as different options are debated. Even if we could provide some robust and clear definition of self-interest—and furthermore argue that it is indeed (in most cases) preferable to other considerations—privileging one's self-interest would only work at the level of individual behaviour and interactions.

There are economic models that assume that individuals usually act in their self-interest. In these models, when other individuals also act in their self-interest, this can produce unintended benefits. Individual shoppers in a supermarket choose the shortest and the fastest-moving queue as it is in their self-interest to do so, and when others choose similarly, you get roughly equal queues that benefit all customers. However, for self-interest to lead to the most efficient outcome (even queues and reasonable check-out times), the customers must have perfect knowledge (they need to see all the queues), and there should be no disrupting factors involved (e.g., someone holding up the queue to return an item).[15] This is the applicability of self-interest in economic theory. It is quite a leap from this to argue that policy decisions can be based on

[14] The hammer-example is also deeply flawed as an example of self-interest in free-market liberalism. Rather than say "Stop hitting yourself with a hammer", many free-market advocators would probably say that it is permissible to hit yourself with a hammer, and barring extraordinary circumstances, others are not allowed to interfere if their persuasion has been unsuccessful.

[15] Thank you to N. Emrah Aydinonat for suggesting this example.

self-interest alone without the need for ethical arguments, especially since in commons problems single-minded pursuit of individual self-interest leads to game-theoretical problems.[16]

In any case, short-term and long-term self-interest are often in conflict. This is the third problem with the pure self-interest approach. Weisbach seems to acknowledge as much, as in a footnote he reveals that his own view is that the United States should have ratified the Kyoto Protocol, despite its problematic features (he describes the treaty's design as 'terribly flawed'):

> Starting fuel switching sooner rather than later is, I believe, cost saving, and ratifying the Kyoto Protocol might have accelerated this process in the United States. If we eventually have to move to renewable energy, we might as well start now and do it slowly rather than wait and try to do it all at once. Moreover, it might have made negotiations with developing countries over their commitments easier (although this is highly speculative). By accelerating fuel switching and helping to induce developing countries to agree to emissions reductions, ratifying the Kyoto Protocol might well have been in the United States's self-interest. (Weisbach 2016: 169)

The last line of the quote is revealing: ratifying the treaty *might well have been* in the United States's self-interest. There is rarely a simple answer to what is in someone's self-interest, especially when our discussion is at the level of collective actors and policy decisions. We must weigh the pros and cons, consider how the policy is likely to affect different interest groups, to what degree, and so on. While I do not claim that any of this would come as a surprise to Weisbach, the complexity of self-interest seems to elude him, as he does not problematise the almost king-maker quality he affords it in decision-making. Self-interest as a default preference is not a foregone conclusion. Not just that, but capturing our preferences and acting accordingly is a more complex procedure than trying to work out what is in fact in our self-interest. This is because we can value things that are not in our self-interest, as I argued in the previous section.

[16] Even though climate change is often characterised in terms of free-riding concerns, I am not convinced that we are dealing with a Prisoner's Dilemma (see Aklin and Mildenberger 2020).

Climate change has become a problem that has a strong self-interest aspect only when we have perhaps left it too late. Basing policy on pure self-interest represents false technocratic neutrality and is too simplistic. While it is clear that we need to act fast and that doing so is also now in our self-interest, *how* we should act is less clear. There are many solutions and many options for reducing emissions, and deliberation and debate are needed to determine what to prioritise and how to implement the changes in a given social context. Ethical arguments by philosophers and other normative theorists are part of this wider debate over what are the best policies. Philosophers can contribute to the public debate in the traditional style of writing papers and giving talks, or they can engage in debates in more direct ways through interactive participation, where the debate affects the content of the normative constructs that the theorist puts together (Green and Brandstedt 2021). Regardless of the approach taken, such arguments are contributions to public debate, attempts to think and argue as precisely and clearly as possible, making use of philosophy's long tradition in this. Philosophical arguments are meant to be argued over and discussed, to stand on their own two feet and be tested in debates, not something to be accepted and adopted wholesale without thinking them through first. (This does not mean that there are no ethical truths, see Chap. 1.)

This is not the function Weisbach (2016, 153) wants to assign to them, as he interprets philosophical arguments as highly prescriptive, listing them along with things like religion and family, which might have an influence on our thinking. Weisbach's somewhat dogmatic view of ethics is illustrated most clearly in the following quote (Gardiner and Weisbach 2016, 257):

> The task for those defending the use of ethics to shape climate change policy is to put forward a theory of ethics that does not suffer from internal logical problems, that can actually be implemented, and that when implemented will solve the problem. This is not a high bar. It simply asks that if we are to follow a theory, that it makes sense and work.

But this is setting a high bar. While ethics should not suffer from problems of internal logic, it need not be about coming up with a theory of what is right that can then be straightforwardly implemented in a society.[17] This would be setting an unfeasible bar, as it confuses the function of normative ethical arguments with the tools that are needed to come up with decisions about how we should live together in societies (see Chap. 1). Instead, philosophical ethics can be conceptualised as discussing normative concerns in an open and critical manner, about looking deeper into our values and challenging our assumptions. Philosophical theories, such as those on how climate burden-sharing should be arranged, are important contributions to the wider social and political debates. But just like valuable economic models or other inputs, they on their own cannot be the final and only word in deciding how we organise our lives together locally and globally. Discussions on ethical principles are valuable and necessary for well-designed climate policies, but not sufficient on their own, as the policy context has to be assessed with all its complexities and uncertainties (Kowarsch and Edenhofer 2016). In any case, philosophical theories are meant to be debated and examined. Or as Mary Geach (2005, xvii) puts it: "The difference between a philosopher [...] and a sage who is giving of his wisdom, is in this: that the way to show respect for a sage is to accept his teaching, but the way to respect the philosopher is to argue." Ethical theories can highlight what is at stake in public debates and they can help ground normative arguments, but they are not meant to be taken as gospel, and nor do they need to claim to offer comprehensive policy proposals that can be adopted wholesale. That is not what they are about: they are about the values that are at stake, whereas democracy is about trying to find the right balance in decisions that affect us all. Engaging with ethical theories can also lead to ethical learning, as when we really think hard about an issue, especially together in deliberation, we are bound to have new insights.

[17] Although he frames his response through the perfect moral storm metaphor, Gardiner (2016, 46–47) also criticises the 'Chicago lawyers' for trying to frame climate change as a 'pure policy' problem in which ethics are not required. He disparages how "those with ethical concerns—philosophers, political leaders, civil society groups or members of humanity at large—are told to cede the floor to experts in the favored disciplines", such as economics and international relations (ibid., 47). "Put bluntly, rather than confronting us all as the great moral and political challenge of our time, climate change becomes a matter best left to the technocrats." I agree with Gardiner that Chicago welfarism is an ethical doctrine, and a complex and controversial one to boot (ibid., 81). However, this objection is somewhat sidelined by the focus on the perfect moral storm, especially the theoretical storm, allowing Weisbach to conclude that he agrees with Gardiner "on the central issue: existing theories of ethics fail" (Gardiner and Weisbach 2016, 259).

Concluding Remarks

I have discussed the impossibility of neutral technocratic climate policy by examining the inherent normativity of economics and the complexity of national self-interest. Climate policy is about values, not just numbers, and not all value choices are subject to optimisation. The problem is not that existing ethical theories fail when faced with the climate change challenge. The problem is the existing systems and structures often do not allow for enough public debate of the ethical dilemmas involved and the normative choices at stake. In general, public value choices should be debated through participatory and deliberative methods. Normative arguments by philosophers have an important role to play as part of this bigger debate.

We might still have time to prevent an all-out catastrophe if we act right now. However, there is a lot of uncertainty and the science becomes more alarming with every passing year that is wasted. This urgency might mean that we need some purely technocratic quick decisions in addition to participatory debate to get the most urgent mitigation and adaptation actions underway. Even so, we should make sure that public awareness of the normative choices is heightened and the role of genuine democratic debate increases as we continue the process and improve our democratic institutions. What is not in doubt is that if we can still manage to avoid a global catastrophe, we must learn from past mistakes. With the climate crisis we have waited until a disaster is on our doorstep.

I have argued throughout the book that dividing climate change responsibilities sharply into individual and collective ones presents a false dichotomy. The challenge climate change brings with it is inherently collective and structural, but individuals are the embedded creators of these structures. Solving the climate puzzle means revamping infrastructure, consumption norms, and institutions. It is a big task to be sure. Still, human beings are innovative and social creatures. When we start viewing our world from slightly different angles, we may begin to see interconnectedness and new possibilities.

References

Aklin, Michaël, and Matto Mildenberger. 2020. Prisoners of the Wrong Dilemma: Why Distributive Conflict, Not Collective Action, Characterizes the Politics of Climate Change. *Global Environmental Politics* 20 (4): 4–27.

Anderson, Elizabeth. 2003. Sen, Ethics, and Democracy. *Feminist Economics* 9 (2–3): 239–261.

Beeson, Mark. 2010. The Coming of Environmental Authoritarianism. *Environmental Politics* 19 (2): 276–294.

Benson, Jonathan. 2019. DELIBERATIVE democracy and the Problem of Tacit Knowledge. *Politics, Philosophy & Economics* 18 (1): 76–97.

Blue, Gwendolyn. 2017. Participatory and Deliberative Approaches to Climate Change. *Oxford Research Encyclopedia of Climate Science*, March 29.

Brennan, Jason. 2016. *Against Democracy*. Princeton: Princeton University Press.

Broome, John. 1992. *Counting the Cost of Global Warming*. Cambridge: White Horse Press.

Hatfield, Chisholm, Elizabeth Marino Samantha, Kyle Powys Whyte, Kathie D. Dello, and Philip W. Mote. 2018. Indian Time: Time, Seasonality, and Culture in Traditional Ecological Knowledge of Climate Change. *Ecological Processes* 7: 25.

Coleman, Martha, Laura Devaney, Diarmuid Torney, and Pat Brereton. 2019. Ireland's World-Leading Citizens' Climate Assembly. What Worked? What Didn't? *Climate Change News*, June 27.

Davidson, Marc D. 2017. The Ethics of Discounting: An Introduction. In *The Ethical Underpinnings of Climate Economics*, ed. Adrian Walsh, Säde Hormio, and Duncan Purves, 22–40. London: Routledge.

Dryzek, John S., and Simon Niemeyer. 2019. Deliberative Democracy and Climate Governance. *Nature Human Behaviour* 3 (5): 411–413.

Fischer, Frank. 2017. *Climate Crisis and the Democratic Prospect: Participatory Governance in Sustainable Communities*. New York: Oxford University Press.

Gardiner, Stephen M. 2016. In Defense of Climate Ethics. In *Debating Climate Ethics*, ed. Stephen M. Gardiner and David A. Weisbach, 3–133. New York: Oxford University Press.

Gardiner, Stephen M., and David A. Weisbach. 2016. *Debating Climate Ethics*. New York: Oxford University Press.

Geach, Mary. 2005. Introduction. In *Human Life, Action, and Ethics: Essays by G.E.M. Anscombe*, ed. Mary Geach and Luke Gormally, xiii–xxi. St Andrews Studies in Philosophy and Public Affairs. Exeter: Imprint Academic.

Goulder, Lawrence H., and Roberton C. Williams III. 2012. The Choice of Discount Rate for Climate Change Policy Evaluation. *Climate Change Economics* 3 (4): 1250024.

Green, Fergus, and Eric Brandstedt. 2021. Engaged Climate Ethics. *The Journal of Political Philosophy* 29 (4): 539–563.

Hickman, Leo. 2010. James Lovelock: Humans Are Too Stupid to Prevent Climate Change. *The Guardian*, March 29.

Hormio, Säde. 2017. Climate Change Mitigation, Sustainability and Non-substitutability. In *The Ethical Underpinnings of Climate Economics*, ed. Adrian Walsh, Säde Hormio, and Duncan Purves, 103–121. London: Routledge.

Huber, Robert A. 2020. The Role of Populist Attitudes in Explaining Climate Change Skepticism and Support for Environmental Protection. *Environmental Politics* 29 (6): 959–982.

Jamieson, Dale. 2021. Marooned Democracy: Climate Change Demands Management, Not Solutionism. *Georgetown Journal of International Affairs*, February 5. https://gjia.georgetown.edu/2021/02/05/marooned-democracy-climate-change-demands-management-not-solutionism.

Kelman, Steven. 1981. Cost-Benefit Analysis: An Ethical Critique. *AEI Journal on Government and Society Regulation* February: 33–40.

Kowarsch, Martin, and Ottmar Edenhofer. 2016. Principles or Pathways? Improving the Contribution of Philosophical Ethics to Climate Policy. In *Climate Justice in a Non-Ideal World*, ed. Clare Heyward and Dominic Roser, 296–318. Oxford: Oxford University Press.

Landemore, Hélène. 2013. *Democratic Reason: Politics, Collective Intelligence, and the Rule of the Many*. Princeton: Princeton University Press.

Lepori, Matthew. 2019. Towards a New Ecological Democracy: A Critical Evaluation of the Deliberation Paradigm Within Green Political Theory. *Environmental Values* 28: 75–99.

Lövbrand, Eva, Mattias Hjerpe, and Björn-Ola Linné. 2017. Making Climate Governance Global: How UN Climate Summitry Comes to Matter in a Complex Climate Regime. *Environmental Politics* 26 (4): 580–599.

MacKenzie, Michael K., Maija Setälä, and Simo Kyllönen. 2023. *Democracy and the Future: Future-Regarding Governance in Democratic Systems*. Edinburgh: Edinburgh University Press.

Marshall, George. 2014. *Don't Even Think About It: Why Our Brains Are Wired to Ignore Climate Change*. New York: Bloomsbury.

Mittler, Daniel. 2014. The Changing Ethics of Climate Change. *Ethics & International Affairs* 28 (3): 351–358.

Moshman, David, and Molly Geil. 1998. Collaborative Reasoning: Evidence for Collective Rationality. *Thinking and Reasoning* 4 (3): 231–248.

National Research Council. 1983. *Changing Climate: Report of the Carbon Dioxide Assessment Committee*. Washington, DC: The National Academies Press. https://doi.org/10.17226/18714.

Neimark, Benjamin, John Childs, Andrea J. Nightingale, Connor Joseph Cavanagh, Sian Sullivan, Tor A. Benjaminsen, Simon Batterbury, Stasja Koot, and Wendy Harcourt. 2019. Speaking Power to "Post-Truth": Critical Political Ecology and the New Authoritarianism. *Annals of the American Association of Geographers* 109 (2): 613–623.

Niemeyer, Simon. 2020. Deliberation and Ecological Democracy: From Citizen to Global System. *Journal of Environmental Policy & Planning* 22 (1): 16–29.
Nightingale, Andrea Joslyn, Siri Eriksen, Marcus Taylor, Timothy Forsyth, Mark Pelling, Andrew Newsham, Emily Boyd, Katrina Brown, Blane Harvey, Lindsey Jones, Rachel Bezner Kerr, Lyla Mehta, Lars Otto Naess, David Ockwell, Ian Scoones, Thomas Tanner, and Stephen Whitfield. 2020. Beyond Technical Fixes: Climate Solutions and the Great Derangement. *Climate and Development* 12 (4): 343–352.
Nordhaus, William. 2007. A Review of the Stern Review on the Economics of Climate Change. *Journal of Economic Literature* XLV: 686–702.
Ober, Josiah. 2008. *Democracy and Knowledge: Innovation and Learning in Classical Athens*. Princeton: Princeton University Press.
Oreskes, Naomi, and Erik M. Conway. 2010. *Merchants of Doubt: How a Handful of Scientists Obscured the Truth on Issues from Tobacco Smoke to Global Warming*. New York: Bloomsbury Press.
Posner, Eric A., and David Weisbach. 2010. *Climate Change Justice*. Princeton: Princeton University Press.
Schmidt, Robert C. 2021. Are There Similarities Between the Corona and the Climate Crisis? *Journal of Environmental Studies and Sciences* 11: 159–163.
Schroeder, Heike. 2010. Agency in International Climate Negotiations: The Case of Indigenous Peoples and Avoided Deforestation. *International Environmental Agreements: Politics, Law and Economics* 10 (4): 317–332.
Schulz-Hardt, Stefan, Marc Jochims, and Dieter Frey. 2002. Productive Conflict in Group Decision Making: Genuine and Contrived Dissent as Strategies to Counteract Biased Information Seeking. *Organizational Behavior and Human Decision Processes* 88 (2): 563–586.
Schwartzberg, Melissa. 2015. Epistemic Democracy and Its Challenges. *Annual Review of Political Science* 18: 187–203.
Shearman, David, and Joseph W. Smith. 2007. *The Climate Change Challenge and the Failure of Democracy*. Praeger.
Sen, Amartya. 2007. Climate Policy as Human Development. Special contribution in *Human Development Report 2007/2008. Fighting climate Change: Human Solidarity in a Divided World*, Kevin Watkins et al., 28–29. United Nations Development Programme. Basingstoke: Palgrave Macmillan.
Shue, Henry. 1999. Global Environment and International Inequality. *International Affairs* 75: 531–545.
Singer, Peter. 2002. *Global Ethics: One World: The Ethics of Globalization*. New Haven: Yale University Press.
Stern, Nicholas. 2007. *The Economics of Climate Change: The Stern Review*. Cambridge: Cambridge University Press.
Stevenson, Hayley. 2013. *Institutionalizing Unsustainability: The Paradox of Global Climate Governance*. Berkeley: University of California Press.

Stevenson, Hayley, and John S. Dryzek. 2014. *Democratizing Global Climate Governance*. Cambridge: Cambridge University Press.
Tabuchi, Hiroko. 2017. As Beijing Joins Climate Fight, Chinese Companies Build Coal Plants. *The New York Times*, July 1.
Vanderheiden, Steven. 2008. *Atmospheric Justice: A Political Theory of Climate Change*. New York: Oxford University Press.
Weisbach, David A. 2016. The Problems with Climate Ethics. In *Debating Climate Ethics*, ed. Stephen M. Gardiner and David A. Weisbach, 137–244. New York: Oxford University Press.
Weitzman, Martin L. 2011. Fat-Tailed Uncertainty in the Economics of Catastrophic Climate Change. *Review of Environmental Economics and Policy* 5 (2): 275–292.
Willis, Rebecca, Nicole Curato, and Graham Smith. 2022. Deliberative Democracy and the Climate Crisis. *Wiley Interdisciplinary Reviews: Climate Change*, e759.

Index[1]

[1] Note: Page numbers followed by 'n' refer to notes.

S. Hormio, *Taking Responsibility for Climate Change*,
https://doi.org/10.1007/978-3-031-51753-2

The manufacturer's authorised representative in the EU is Springer Nature Customer Service Centre GmbH, Europaplatz 3, 69115 Heidelberg, Germany. If you have any concerns regarding our products, please contact ProductSafety@springernature.com

Printed and bound by CPI Group (UK) Ltd, Croydon, CR0 4YY

01/10/2026

02225879-0009